DECENTRALIZING
ELECTRICITY
PRODUCTION

DECENTRALIZING ELECTRICITY PRODUCTION

HOWARD J. BROWN, Editor
with Tom Richard Strumolo

Yale University Press
NEW HAVEN AND LONDON

Published with assistance from the Kingsley Trust Association Publication Fund established by the Scroll and Key Society of Yale College.

Copyright © 1983 by Yale University.
All rights reserved.
This book may not be reproduced, in whole or in part, in any form (beyond that copying permitted by Sections 107 and 108 of the U.S. Copyright Law and except by reviewers for the public press), without written permission from the publishers.

Designed by James J. Johnson
and set in Melior Roman type by
P & M Typesetting, Inc.
Printed in the United States of America by
Halliday Lithograph, West Hanover, Massachusetts.

Library of Congress Cataloging in Publication Data
Main entry under title:

Decentralizing electricity production.

   Includes index.
   1. Electric power production—Addresses, essays, lectures. I. Brown, Howard J., 1945–
II. Strumolo, Tom Richard, 1952–
TK1005.D38   1983   363.6'2   83-3677
ISBN 0–300–02569–6

To R. Buckminster Fuller, whose ideas, lifetime work, and belief in individual initiative provide the inspiration to "dare to be naive."

Contents

Preface and Acknowledgments	ix
Part I: The Context for Reform	1
1. Problems, Planning, and Possibilities for the Electric Utility Industry Howard J. Brown	3
2. Technology Is the Answer! (But What Was the Question?) Amory B. Lovins	19
3. The Pendulum Swings Again: A Century of Urban Electric Systems David Morris	37
Part II: Managing Decentralization: The Tools, Rules, and Fuels of Reform	59
4. The Potential for Diversity: The Production Alternative Lisa Frantzis	61
5. The Electrical Energy Production System in Transition: The Critical Factor of Reliability Robert D. Morris	93
6. The Grid as Energy Absorber and Redistributor Bent Sørensen	109
7. Economic Feasibility of Dispersed Solar Electric Technologies Robert E. Witholder, Jr.	121
8. The Advantages of Integrating Decentralized Renewable Electrical Technologies: A Connecticut Case Study James Gustave Kahn	163
9. Restructuring the Electric Utility Industry: A Modest Proposal David A. Huettner	183
10. Cogeneration and Small Power Production: Some Intergovernmental Policy Concerns Edward Thompson III	199

Part III: Issues in Decentralization — 213

11. *Diseconomies of Scale* — 215
    David A. Huettner

12. *Energy: Jobs and Values* — 229
    James W. Benson

13. *Planning Practically for a Decentralized Electrical System: How Past Experience Can Guide Us* — 243
    E. F. Lindsley

Contributors — 255

Index — 258

Preface and Acknowledgments

This is a book about the future, yet it includes no forecasts or predictions. Our motivation to prepare the book did not result from a curiosity about what is *likely* to happen but from an exploration of what *could* happen. As planners involved in energy issues daily, we began with the belief that energy—the way we produce, distribute, and use it—is at the heart of every social, economic, and environmental problem confronting society; and that energy policies and decisions based on projections and forecasts derived from recent trends can only encourage more of what we are already doing. It seems self-evident that if there is one thing we do not need, it is more of what we already have.

Fossil and nuclear fuel shortages, health and safety problems, inflationary pressures, environmental impacts, and financial problems are threatening the viability of our electrical energy system. Many people who are involved with these issues have come to feel that the problems inherent in the way society produces and distributes electricity are so serious that major changes are in order. But it is the responsibility of citizens who disagree with the status quo to propose alternatives, and that is what this book is about. Nevertheless, we do not presume that it contains answers, only possibilities.

In the numerous debates over nuclear power and over various specific alternative technologies, much of the attention given to alternatives focuses on forecasts of likely penetration into the market based on past trends. The most general question we address in this work is, "If we were to create a decentralized electrical energy system, how would it work?" As planners we concerned ourselves with only two general constraints: what is technologically possible (using existing knowledge and tools) and what is ecologically possible (considering existing conditions). Economic viability is something society creates after deciding what it wants, not something that should determine its needs. Making something economical is a policy problem. The question we need to ask is, "How do we do it?"—not "Should we?"

This outline of alternative methods for producing and distributing electricity clearly raises more questions than it answers, but we think it should be useful to anyone wanting to understand the options open to society. We hope it will serve as the basis for more research, for more thoughtful and meaningful discussion of plans by public and private agencies, and for the establishment of clearer social goals.

Our exploration of the question of alternatives for electrical energy generation began in 1972 at Earth Metabolic Design, Inc. (EMD), a nonprofit research organization concerned with prudent resource planning. At that time Medard Gabel, EMD's director, organized a project to assess the potential of using renewable energy sources on a global scale. The project is described in his book *Energy, Earth and Everyone*. Numerous subsequent research and planning projects conducted by private and government agencies provided opportunities for expanding our thoughts on the topic. Thus, this book is largely the result of ten years of research in alternative energy systems and of a conference entitled, "The Problems and Potentials of Decentralizing Electrical Production," held at Wesleyan University (Middletown, Connecticut) in April 1978. The one-day symposium was sponsored by the College of Science in Society at Wesleyan, and by EMD, Northeast Utilities, the Connecticut State Energy Office, and the New England Regional Office of the Department of Energy. Some of the speakers have contributed articles to this book.

The conference literature set forth the following questions to be addressed:

> Can significant amounts of electricity be produced from small-scale renewable sources such as wind, photovoltaics, cogeneration, and hydroelectric? How much?
> Have these systems proved themselves? Are they reliable?
> Can they compete economically with fossil and nuclear fuels?
> What are appropriate scales of production?
> What kinds of institutional changes would be required?
> What kinds of technological changes would be required?
> Can electricity be produced in competitive markets by small producers?
> Can continuous supply be assured? Can quality be assured?
> How would the system be managed? What role would utilities play?
> Would decentralized systems save money? Would they pollute more or less?
> How would electrical power be stored?

Since the conference we have added some more critical questions:

> What is "appropriate" technology regarding electrical generation?
> To what extent can wind supply this country with electricity? Hydroelectric power? Cogeneration? Photovoltaics? Solid waste?
> Can other sources such as hydrogen and fuel cells supply a sufficient amount of electricity?
> How could a diversified grid be decentralized, integrated, and managed?
> How can past experience guide us?
> Are there diseconomies of scale of electrical production?
> What impacts on employment would a decentralized system have?
> How can such a system be implemented?

PREFACE AND ACKNOWLEDGMENTS

We have taken special care to present a comprehensive view of the argument by dealing with as many aspects as possible—political, economic, legal, environmental, and technical. The conception of an alternative structure for the electrical energy industry that emerges from this volume represents our own views, which are not necessarily those of the individual authors whose work is included. In our own research over the years we evolved the overall thesis of the book and identified the areas of research and expertise that would be required to establish the concept's feasibility. We invited each of the authors because we felt he or she was making a creative contribution in those fields that are important to the future of electrical energy systems. However, each of the authors, if given the responsibility of editing this volume, might draw a very different set of conclusions from the material presented.

The first three articles, comprising Part I, form the philosophical framework for the rest of the book and establish an argument for decentralization. Part II explores how decentralized production in a utility grid could be managed and provided. It contains articles on the resources and technologies of production; the technologies for grid integration and storage; the economics, management (including public policy), and engineering; and the role of the marketplace. Part III includes articles on a range of political, practical, ideological, and economic issues relating to decentralization.

We want to express our appreciation to Constance Ettridge, Gunhild Gross, Daniel Bob, and James Dray for their patient and tireless help. Without their assistance this project would never have been completed.

Howard J. Brown
Tom Richard Strumolo

PART I

The Context for Reform

HOWARD J. BROWN

CHAPTER 1 | *Problems, Planning, and Possibilities for the Electric Utility Industry*

PROBLEMS: A MARGINAL UTILITY

Electric utilities have problems that can be understood only in the larger social and ecological context. More and more of a rapidly growing world population wants access to the resources needed to create prosperity. Yet the resources on which we have become dependent are finite. Given present patterns of using these resources, there simply are not enough to ensure both prosperity and equality; and the biosphere cannot absorb all the by-products of the endless expansion needed to meet global needs. Simultaneous inflationary and recessionary pressures (and the threat of supply disruptions) are among the early signs of stress resulting from this predicament.

Decreasing net productivity from dependence on traditional fuels and political maneuvering for control of remaining reserves both contribute to the problem. Basic supply and demand economics is based on the assumption that as resources become scarce, rising prices from the supply/demand imbalance will pay for finding more. But the concept of net energy availability illustrates the problem with that assumption: the closer we come to the end of our fuel reserves, the more energy it takes to get each unit into usable form.

This logic was described by physicist Gerald Feinberg in his now prophetic 1969 book, *The Prometheus Project*. In the early developmental stages of the American commercial nuclear fission industry, economists recognized the finite supplies of concentrated nuclear fuels in the earth. Since they understood that uranium exists in dispersed form in large quantities (even in common granite), they assumed that as concentrated resources ran out, higher prices would pay for mining less concentrated reserves. Feinberg pointed out what economists did not recognize, that it may take more energy to get that dispersed uranium than could be extracted from it once mined.

Ecologist Howard Odum used the net-energy concept to describe the problem of our dependence on petroleum. Each new barrel of oil is harder to find, deeper in the ground (harder to mine), and lower in quality (harder to refine). Because of this, Odum argued, an inherent relationship exists between dependence on fossil fuels and inflation. As we get closer to the end of concentrated reserves, we find that each new barrel yields less useful net energy; thus, we spend more and more of each dollar on getting energy and less for the desired product. Long before we actually run out of fuel, we may come to the point where we are spending more energy than we are producing. And the theory can be extended: long before we reach the point of negative net productivity, we may reach the point where there may not be enough traditional fuels to build alternative technologies and maintain the fabric of society at the same time—at least, given present productivity.

The inflationary impact of continuing dependence on diminishing resources is amplified by the ever-increasing demand for capital. As Lovins points out in chap. 2, the capital required for finding, extracting, preparing, and safely converting fuels from a shrinking resource base drains the rest of the economy of finite capital. In the process, the development of alternatives is retarded. Thus we are already feeling, and will increasingly feel, the economic impact of dependence on disappearing fossil fuels—it is not just a problem for the future. Ecological limits are now impinging on our economic system; and without structural reform, institutions have difficulty adapting. Political and economic turmoil over the control of petroleum, uranium, and other nonrenewable resources is a logical result of diminishing supplies.

Given the new ecological realities, accepted policy levers are simply not effective. For example, levers traditionally used to combat recession (i.e., stimulate economic growth) only stimulate inflation by increasing demand and competition for finite, harder to get, energy-expensive resources. When prices rise but the net availability of resources needed to meet demand actually declines, both recession and inflation accelerate. Prices rise but production and consumption do not rise proportionately. Conversely, levers used to combat inflation by slowing down the economy also accelerate inflation and recession. When the economy slows down, production declines and more people lose work, but prices do not come down because the energy problem remains. In fact, because capital investment in scarce fuels only rises as a share of total economic activity, prices only rise further. Inflation and recession can be interpreted as two sides of the same coin—feedback from a finite biosphere telling us that present patterns of expansion cannot continue indefinitely. For the first time, argues economic activist Hazel Henderson in her book *Creating Alternative Futures* (New York: Berkley, 1978), our economy is being driven and directed primarily by ecological forces rather than by policymakers.

The perspective of a finite world, however, does not lead irrevocably to the assumption that there are limits to progress and wealth. Evidence is accumulating that with proper management, available resources may well prove adequate for long-term prosperity and equality. It is apparent, however, that if improperly managed, our resources may become scarce, leading to further economic, political, and ecological deterioration. Thus, society's concern over access to resources is caused not by their scarcity or by ecological limitations but by the way we think about ourselves and our problems and the way we manage (or fail to plan and manage) the resources available.

Planning and management decisions are based on assumptions about wealth, human behavior, and the environment. When these ideological and economic assumptions inaccurately describe our place in the world (i.e., the desirability of undirected economic growth), they exacerbate the problems.

There seem to be universal identifiable stages in the way systems relate to their environment. For example, Howard Odum points out that "during times when there are opportunities to expand energy inflows, the survival premium by Lotka's principle is on rapid growth, even though there may be waste."[1] This condition describes the period of rapid growth evident both in the early development of ecosystems and in the growth of Western industrial economies to the present. These periods of rapid growth, Odum maintains, require the expansion of resource availability.

> We observe dog-eat-dog growth competition every time a new vegetation colonizes a bare field in rapid expansion to cover the available energy receiving surfaces. The early growth ecosystems put out weeds of poor structure and quality, which are wasteful in their energy-capturing efficiencies, but effective in getting growth even though the structures are not long lasting.
>
> Most recently, modern communities of man have experienced two hundred years of colonizing growth, expanding to new energy sources such as fossil fuels, new agricultural lands, and other special energy sources. Western culture and more recently, Eastern and Third World cultures, are locked into a mode of belief in growth as necessary to survival. "Grow or perish" is what Lotka's principle requires but only during periods when there are energy sources that are not yet tapped.

Odum points out that during times when energy sources have been depleted and new sources cannot be found, Lotka's principle requires "that those systems win that do not attempt fruitless growth but instead use all available energies in long-staying, high-diversity, steady-state works." In this condition, which prevails when supplies of critical resources are dwindling, *survival is dependent on the type of resources we use and how well we use them rather than on how much or how many we use.*

The explosion of interest in biology and ecology during recent decades has dramatically improved science's understanding of how living systems behave and adapt successfully under differing conditions. Yet we have not effectively integrated this knowledge into the study, planning, and management of our social and technological systems. Nowhere is the gulf between new realities and old perceptions more evident than in our electrical energy production and distribution system.

The electric utility industry of the United States can be characterized as a vertically integrated industry composed mostly of publicly regulated and protected regional monopolies. These institutions are assigned the right and responsibility to produce and distribute sufficient, reliable, and high quality electricity to meet consumer demand in the most economical fashion. Many electrical utilities are finding it difficult to meet these requirements and remain solvent.

Not long ago private electric utilities represented ideal security for investors, but in recent years utilities have become questionable investments in many parts of the country. Private utilities, like other corporations, must maintain or increase their income to remain attractive investments to stockholders, and there are three primary ways in which they can accomplish this: sell more electricity, raise prices, or reduce production costs. Each of these options is discussed below.

Historically, increasing sales has been the simplest way for utilities to raise revenues. In periods of widely available energy resources (with which to generate electricity) and rapidly expanding economic activity, utilities fit neatly into the Keynesian economic model. The best way to stimulate growth was to build new generating capacity, thereby increasing the availability of low-cost electricity to sell; that low-cost electricity, in turn, helped attract new growth which would consume more electricity, raising revenues and permitting further construction. This cycle of growth made the interests of the utilities seem allied directly with the general interests of the people of a region, state, or country because of the accepted Keynesian assumption that new business activity means more jobs, more spendable income, further business activity, and so on. Until recently, limits to this cycle of growth were not considered by utility planners or state and federal regulators. States began and continue to allow utilities to reflect capital investment in rate structures, thus encouraging them to expand. This rate-base accommodation in conjunction with the long-term federal tax advantages of increased capital investment made expansion seem very logical.

Thus, under advantageous economic conditions, it has been widely assumed that a mutually supportive relationship exists between the availability of electricity and economic growth. Utilities borrow money from private investors to build new generation capability, and generation capability in turn stimulates the economy to such a degree that consumption increases and revenues from new customers are sufficient to pay off the loans.

A new cycle has set in, however, in which costs of capital and fuel are rising while the growth of both the population and the economy is slowing down. Therefore, in many parts of the country, utilities are finding themselves in the position of having to pay off past loans for current overcapacity without experiencing a concomitant increase in new sales. As prices of electricity rise, elasticity becomes an increasing factor as customers begin to reduce consumption by conserving, switching to other energy forms, and even producing their own. Thus, increasing sales will be less and less of an option for generating the income to pay off loans and increase future sales, and limited cash flow will become an increasingly serious problem.

Another way that utilities, like other businesses, can raise revenues is by increasing the price of the commodity produced. Since most utilities are monopolies, they are subject to the regulatory control of local, state, and federal agencies; but until recently utilities have had little difficulty in gaining the approval of public utility commissions and other regulatory agencies for rate hikes. Because the cost of producing electricity has been very low in relation to the society's overall standard of living, because electricity is a relatively clean and flexible form of energy, and because it is essential for many end uses, price elasticity has been low and price increases have not significantly affected demand.

If profits from increasing sales are not available to utilities as new sources of income, neither are increasing rates because the inflationary pressures that have brought about the spreading consumer movement have placed new pressures on public utility commissions to scrutinize closely utility requests for rate increases and new facilities. Lower profits mean higher interest rates and further rising costs.

The last mechanism for increasing income is to reduce costs by, for example, producing energy from newer, cheaper sources, improving management techniques, and improving the efficiency of transmission. To understand the seriousness of the dilemma confronting the utilities, we must reexamine, in the light of new circumstances, the three mechanisms utilities have historically used to increase profits.

Increasing fuel and capital costs, environmental and safety concerns, as well as general inflation are driving up the cost of electrical generation. Reducing costs is simply not an option available to public utilities, whose mandate is to provide reliable electric power and who come under severe criticism from the public and from decision makers when reliability is sacrificed. Costs have remained as low as they have in the nuclear and fossil fuel-based utility industry because of the large degree to which total costs have been externalized—borne by government or (in the case of nuclear) simply left for future generations. But increasing demands for institutional responsibility are likely to force increased internalization of costs now externalized, causing further major increases in costs for utilities and a further erosion of sales potential.

As conservation is accomplished, the high fixed costs from nuclear investments mean that unit costs for consumers rise further, encouraging more conservation and small-scale alternatives. Thus the utilities, fulfilling their social mandate to produce more electricity, are caught in the economic dilemma. Many have made enormous commitments to new nuclear power plants, borrowing at ever higher interest rates to pay the spiraling capital costs of new nuclear plants that will produce electricity to stimulate economic growth (and eventually demand) that is simply not likely to occur.

In periods of slow or even nonexistent economic and population growth, even the most traditional economic analysis cannot explain decisions to build. Having already borrowed large sums for plants under construction, utilities must continue to build or be faced with debts often large enough to create fiscal insolvency. In most states, plants that do not produce cannot be included in the rate base. The remaining defense for building the plants is to decrease dependence on foreign oil, but conservation is a cheaper and more effective way to accomplish that goal and one that citizens are likely to carry out on their own as prices rise.

Like many municipal governments, many private utilities are managing money for short-term solvency while selling out long-term stability. When other future costs like those of nuclear waste disposal, plant decommissioning, rising fuel costs, and increasing environmental safeguards are combined with the costs of capital and compared with the economics of conservation, the increasing availability of user-owned renewable capacity, the eroding of the individual's real wealth, and the decreased population (in many areas), it is difficult to understand the logic of advocating further investment in massive nuclear capacity. Major financial institutions are increasingly recognizing this, which is why the cost of money is rising for nuclear plants even faster than in the economy as a whole. Ultimately, someone will have to pay for the future costs, and it will be not only the investors in utilities but the public as well.

So electric utilities, once the symbols of American economic strength, are becoming the symbols and victims of the changing environment—clinging to old perceptions despite new realities. Utility planning, once a primary force in directing the economy, is now being driven by forces outside the economy.

PLANNING: AN ALTERNATIVE TO FORECASTING

When an electric utility goes before a public utilities commission for permission to build additional central power station generation capacity, it relies on forecasts of demand to demonstrate the need and validate its requests. Forecasting is a technique for predicting the future and is used by practically all corporate and government planners in the United States as the basis for formulating national, state, and local policy. It consists of a set of techniques of varying degrees of sophistication and detail but all relying fundamentally on the charting and extrapolation of historical trends to predict future conditions.

A "good" forecast is based on actual monitoring of many social variables related to the subject of a forecast over a long period of time and on the construction of equations describing the relationships. A "poor" forecast tracks only one or a few variables for a short period of time. Most utility forecasts are extremely complex and sophisticated mathematical models designed to describe the quantity and characteristics of electrical energy demand for a region. The significance of utility forecasting in shaping society's future is little understood; and the shortcomings of the methods and assumptions used in forecasting are even less understood.

Many regulators are impressed with the detailed mathematical analysis used in forecast reports. The sheer volume of quantitative material in a forecast lends it a certain scientific credibility with both regulators and the public. Acceptance of the reliability of a utility forecast of demand implies some tacit acceptance of plans for meeting the demand. The character of the supply system, in turn, has a dramatic impact on the future of local, regional, and even national economies (see chap. 2); on the environment; and on the political structure of society. Thus, in an indirect but real way a forecast serves as a plan, but because it is not called a plan many essential responsibilities associated with planning are omitted from the decision-making process.

Two general characteristics of all forecasting must be understood. First, forecasts are constructed on a foundation of values held by the forecaster, the sponsoring institution, the professional community, and the society at large; second, all forecasts, no matter how sophisticated, are based on past patterns of behavior. Both characteristics are discussed below.

Forecasts are constructed on a foundation of value judgments that are almost always implicit. Often the values are not clear, even to the forecaster; sometimes they are intentionally obscured by the masses of data and equations that are used. Each set of data reflects assumptions that led to its selection, and each equation is based on assumptions about behavior, past and future.

Because of the common belief that numbers are inherently objective and because many people are intimidated by advanced mathematics, complex forecasts may actually receive less scrutiny than simple ones. But the values that go into the forecast process very much determine the conclusions that come out. In the end, utility forecasts usually say what forecasters or the employers of forecasters want or expect them to say. Utility models generally support the need for new nuclear or other central station capacities but rarely reflect the need to conserve, decentralize, diversify, or nationalize production. Forecasts to support these latter needs are equally complex, sophisticated, and technically competent, but they are likely to be generated only from models built by individuals whose values and beliefs differ from those of the authors of utility models.

A series of fundamental assumptions are implicit in the demand forecasts used by most utilities, and they should be made explicit. They are:

Economic growth can and should be sustained into the foreseeable future.
A healthy economy requires increasing demand for electricity.
An increased supply of electricity fosters economic growth.
There are sufficient resources to sustain continued economic expansion.
There are no particular ecological limitations to sustained growth.
No particular social or political circumstances will inhibit growth or interrupt supply.
Descriptive mathematical models can be objective.
The future can be predicted by such models with a considerable degree of accuracy.

In addition to these, many other assumptions and values are specific to each forecast or model. The selection of variables to be included in a model is based on assumptions about what is most important, whose data are most accurate, and how far back in time data should be assessed to identify trends and relationships.

The way data are used in the model represents a second layer of value decisions. The equations within the model describe the relationships among the components (or variables). In some cases the equations result from extensive study of the behavior of the variables and their relationships over time; in others the relationships are derived from surveys of "experts" in the field; in yet others they are based on "best guesses" by forecasters. In even the most thoroughly researched examples, values play a role not only in data selection but also in the definition of the relative importance of the variables and in assumptions of causality in the relationships. For example, research may show that over a ten-year period, demand correlated directly with increases in square feet of retail commercial space in a region. Square feet of retail space may then become a reliable indicator of commercial electricity demand. Accepting the relationship, however, requires making assumptions, and the assumptions any individual is willing to make are based on values. Important questions should be asked, for example, "Would the mathematical relationship still be correct if the variables had been plotted and compared twelve, fifteen, or twenty years ago?" "Are the data accurate?" and "Is there any reason to believe that because this relationship held true for ten years it will continue to be true in the future?" The willingness of any forecaster to accept such a relationship as valid is likely to be tied to his predisposition about the outcome.

In many models, variables forecast by in-depth research and sophisticated techniques are used in combination with variables projected by pure guesswork. The research-derived forecasts obscure not only the role of values within them, such as in the example above, but also the more overt value judgments around them. Thus, values are important because they affect the outcome of the forecast; and the forecast, in turn, is used to make policy about the way limited resources are invested. The way they are invested determines (by limiting or encouraging possibilities) directions for the future.

The forecast thus plays a major role in formulating social policy, but the forecaster escapes responsibility for making social choices. As "social scientists," most forecasters view themselves as students of society rather than interveners. The insidiousness of the forecast-based decision-making process clearly limits social responsibility and social choice, which are crucial to democratic systems.

The second major shortcoming of forecasting stems from its reliance on historical trends. If there is one thing we can say with certainty about the future, it is that it will not resemble the past; and if there is one thing we can learn from the past, it is the probability of the improbable. Yet the only image of the future that can emerge from forecasting is a reflection of the past. In a forecast the improbable always remains improbable. Depending on methods and assumptions, forecasts may differ regarding future rates of change, but alternatives, creativity, chance, and nature are all left out.

Early forecasts were constructed from the linear extension of past trends. Russell Ackoff, chairman of the Department of Social Systems Science at the University of Pennsylvania's Wharton School, has observed that the only value of linear trend extrapolation is to show what cannot happen. Thus, such trends can be used to demonstrate why change is necessary but not to determine the character of the change. Science has found no continuous linear trends, and social phenomena seem no exception. Events and forces external to the subject of the forecast intervene to modify the behavior of all systems.

The development of more sophisticated econometric models in the 1960s and 1970s made forecasting more sensitive to many influences ignored by simpler techniques. Yet even the most advanced econometric models are at best reflections of the past. As I have already pointed out, even these models are based on economic assumptions about wealth, success, and growth that simply do not account for a host of external factors that can rapidly affect human behavior (such as the demand for electricity).

Though econometric modeling is an improvement over linear projections, events and forces that can enormously influence demand cannot be included if they occur randomly, irrationally, or so infrequently that the pattern of their occurrence cannot be monitored or predicted; or if they are more subtle than our tools can monitor. Such events include:

political changes (embargoes, price manipulations, sabotage, strikes)
ecological events (earthquakes, severe weather conditions, and other catastrophes)
absolute limits of resources (and their net energy implications)
changes in values (interest in self-sufficiency, concern about pollution, health, and safety)
technological innovation (advances in alternative technologies).

In defending overly optimistic forecasts in the face of shrinking demand after the first oil embargo in 1974, some utility spokesmen argued that the embargo was an "unpredictable" intervention and therefore could not be fairly used to point out the shortcomings of recent forecasts. Such events may not reflect on the quality of forecasts but they do bring into question their reliability.

In spite of the shortcomings of forecasts, individuals and institutions must allocate some portion of their time to preparing for the future. Forecasters are assigned the responsibility to predict the future so that decision makers can know how to invest resources, to prepare. Some forecasters point to their records as evidence that, at least in the short term, they have often been right. But correctness is only a partially valid defense of the role of forecasting. This is true for two reasons: (1) Because many changes occur slowly over a long period of time, predicting trends can have a reasonable degree of reliability in the short term but none in the long or moderate term. (2) The tendency of all prophecies (especially when made by large institutions with large impacts on society) is to become self-fulfilling. The second tendency is very important.

Society in general, and electric utilities in particular, have finite capital resources to invest in any given activity. When utilities create enormous capital investment programs to construct large fossil and nuclear-fueled central generating capacity, they are shaping the future. Customers must pay a share of the plants whether or not they want or need the electricity that will be generated. Consumption has historically been encouraged by rate structures that reward waste, and alternatives for generation have been discouraged because utilities have controlled and restricted access to the distribution system. Only utilities could be guaranteed a return on investment and could attract government subsidies for generating electricity.

Given these circumstances, the tendency of self-fulfilling prophecies to occur during times of rapid economic growth is easy to see. The forecaster predicts rapid growth, the directors and regulators accept the forecast because it is professional and sophisticated, and finite resources are invested in such a way as to encourage and direct growth. Sometimes the participants have been cognizant of the causal relationship between the forecast and consumption, sometimes not; but regardless, the process is insidious. The tendency of utility forecasts to be self-fulfilling does have its limits (given the factors discussed above), but danger exists when institutions and individuals come to believe in and develop a stake in the pattern of linear development that the tendency fosters. Utilities want the conditions that fostered their growth and success in the past to continue in the future. Relying on forecasts for this reason, in the face of changing ecological and economic conditions, is a major contributor to our energy problem. When used as the basis of policy, forecasts can actually obscure both dangers and opportunities ahead. Forecasts cannot point in new directions; and in a rapidly changing world, overreliance on forecasting can be socially maladaptive.

The differences between forecast-generated plans and a comprehensive planning process lie in the establishment of explicit goals and criteria for measuring progress. A forecast is an effort to *predict* the future; to tell what is *likely* to happen. A plan is an effort to *affect* the future; to determine what a planner would *like* to happen, and a method for making it happen. A forecast is ostensibly objective; a plan is intentionally prescriptive. In a forecast, the forecaster's values and goals are usually included only implicitly; in a plan, values and goals are explicit.

In the United States, social planning is often considered contrary to the notions of democratic institutions. In reality, overreliance on forecasting and the lack of a comprehensive planning process are what conflict with democratic ideals by inhibiting choice. A comprehensive plan can include and even encourage diverse (even seemingly conflicting) values, and it can attempt to accommodate unexpected changes.

Until the late 1970s state and federal legislative and regulatory institutions served only to approve or disapprove utility forecasts and plans, never to actively plan or assume responsibility for planning. But the inadequacy of this approach in accommodating resource shortages, rising capital costs, consumer conservation, and other changing factors has given rise to a new conception of utilities and a demand for structural changes in the way we produce and distribute electricity as well as in the institutions that are responsible for it.

The question that consumers and decision makers alike should not ask about the future is, "What is *likely* to happen?" What must be asked is, "What could and should happen to our electric utility system?" "How do we *want* it to work?" Emerging from such questions should be a set of specific goals based on our best assessment of possible problems, environmental conditions, and successful adaptation. Such goals should focus on our ability to:

- produce sufficient electricity for future needs
- minimize future demand through comprehensive system planning and user education
- be resistant to breakdown and intervention by as many ecological and political contingencies as possible (i.e., it should be flexible—diversified and *redundant* in both production and distribution)
- employ the minimum necessary overcapacity
- use indigenous and renewable resources as much as possible
- produce the most inexpensive electricity (consistent with the above)
- employ cost accounting that is comprehensive, long range, and consistent for evaluating alternatives
- be as responsive as possible to new technologies and techniques
- rely to the greatest extent possible on the marketplace for the production of electricity
- minimize environmental impacts.

The next section summarizes the characteristics of a system that could address such goals.

POSSIBILITIES: AN EMERGING IMAGE OF UTILITIES

I have tried to demonstrate that strong environmental constraints preclude resolving the problems of America's electrical utility industry by traditional approaches, but I have also indicated that these constraints do not totally preclude a resolution. The established conception of electric utilities (and their structure) held by utility planners, regulators, and public policy makers is what impedes effective and adaptive responses.

Rapidly changing social, economic, political, and ecological conditions are mandating adaptive structural reform of the electric utility industry. Whether such structural reform is to be accomplished with minimum negative social impact (i.e., with minimum social costs) is a matter of public policy. Delay will raise the costs. Long-term solutions must emerge from a reconsideration of public needs, from a comprehensive assessment of the problems, and from an inventory of all the available technological and policy alternatives.

This book explores the characteristics and components of an alternative approach. Various articles summarize specific characteristics, possibilities, and problems for such a system. I will summarize here the general characteristics in order to create a framework within which the other articles can be understood.

The emerging electrical utility industry will have ten major characteristics:

1. a major role for cogeneration and renewable energy resources
2. more effective utilization of regionally indigenous resources

3. a greatly expanded role for load and end-use management in relation to supply management
4. a greater degree of redundancy in both generation and transmission resulting from reduced economies of scale
5. a greater diversity of resources and technologies for generation
6. the redefinition of a grid from a simple distribution system to an absorber and redistributor of decentralized production
7. a greater dependence on the marketplace to determine the cost effectiveness of various production alternatives
8. a greater degree of institutional separation between the generation and distribution functions of electric utilities
9. the development of an independent service industry to manage decentralized production facilities
10. some expansion of both decentralized and central system storage

A brief introduction to each of these concepts follows.

A Dominant Role for Cogeneration and Renewable Energy Resources
There is heated debate among authorities over how much of future demand can be met by renewable (sustainable and free) energy resources. Utilities have been reluctant to invest in renewable technologies because their reliability in modern electrical grids is largely unproven and because many inherently require small-scale production facilities and, consequently, structural reform to develop. But in the medium-term future, increased use of renewables offers the best practical alternative to rising prices, decreasing net availability, and potential supply disruptions resulting from diminishing reserves of traditional fuels. Further, they offer the hope of increased environmental quality and safety. As safety and environmental costs are increasingly factored into economic assessments, and as prices of traditional fuels rise, the economic attractiveness of alternatives improves.

Chapter 4, by Lisa Frantzis, addresses this subject. It summarizes a series of major independent technology assessment studies to indicate the potential of cogeneration and renewables. It concludes, as have numerous subsequent national studies by the Harvard Business School, the Union of Concerned Scientists, the World Game, and the Solar Energy Research Institute, that renewables could play a major role in our electric energy supply. But this conclusion raises a series of more complex questions about how a system must be organized to use renewables efficiently.

More Effective Utilization of Indigenous Resources
It is important not only that renewable resources be used whenever possible but that those resources that are indigenous to each region be integrated into the system. Resource mappings, such as those used in chapters 4 and 7, indicate that each renewable resource is unevenly distributed geographically but that every region benefits from the availability of some renewables.

Cogeneration (the sequential production of electricity and steam for heat or industrial processes) can be considered an indigenous energy source for the purposes of this discussion. This is because any intensive user of fossil fuels for steam or heat generation is also a potential producer of relatively low-cost electricity. The efficiencies achievable by combining electricity production with such uses are enormous in comparison with separate production. Until recently, only the very largest of industrial users were seen as potential cogeneration sites. Recent advances in cogeneration technology, however, are dramatically reducing the economies of scale. Relatively small-scale systems (under 200 kW) are now being installed throughout the United States at hospitals, universities, public institutions, and at commercial and industrial sites. Very small-scale systems, developed in Israel and in Europe, are now being marketed even for multifamily residential applications. Wastes (municipal, industrial) are other indigenous energy sources discussed in chapter 4. Utilizing indigenous resources increases local self-reliance, reduces transportation costs and energy requirements, and creates economic development (see chaps. 2, 3, and 12).

Expanded Role for End-Use and Load Management
Historically, utilities have planned supply to meet forecasted demand. As supplies diminish and prices rise, managing demand to match available supply becomes a cost-effective activity. This will be true particularly when there is increased reliance on the use of renewable resources because production is more a function of variable conditions than of demand response. The degree of reliability that can be derived from these variable sources is discussed in chapters 5 and 6 and modeled by James Kahn in chapter 8.

As interest in load management has grown, many sophisticated techniques have emerged to control use. They fall into three general categories: (a) those that reduce demand by increasing appliance efficiency and reducing waste; (b) those that direct and control load to make the character of demand curves match the character of supply (i.e., time of day rate, load shedding, interruptable rates, radio control systems); and (c) end-use management (discussed by Lovins in chap. 2). End-use management means matching available energy resource types and characteristics to use requirements and characteristics. Demand reduction and load management represent an entirely new set of planning and management tools that have only begun to be used. They open up new opportunities for increasing reliance on renewable resources.

A Greater Degree of Redundancy in Generation and Transmission
Most renewable resources are distributed—that is, they have relatively lower intensity and are dispersed over large geographic areas, rather than concentrated (as fossil and nuclear fuels are) in relatively few locations. Smaller-scale technologies are required to optimize their conversion to electricity, in comparison with traditional fuels. This means that economies of scale in production must be shifted from the actual generation of electricity to the production of generators.

Mass production of small and medium-scale generators has numerous benefits. A large number of small plants means greater employment per unit of energy (see chap. 12), shorter construction lags (see chap. 2), greater resistance to system disruption (it is less likely that one hundred small plants will go out simultaneously than a single large plant), and reduced demand for capital expenditures for generation because of reduced demand for reserve capacity (see chaps. 2, 6, and 8).

It is important to note that increased redundancy and reduced scale of production units mean greater reliability from renewables than can be expected from single renewable units. It is likely that the wind will be blowing somewhere in a service area all of the time but unlikely that it will be blowing in any single location as often. This is true of most decentralized sources and is discussed in detail by Kahn in chapter 8.

Reconceptualizing the Grid as an Energy Absorber and Redistributor Rather than as a Single Distribution System

Reliance on centralized generation requires a grid in which energy flows one way (conceptually if not literally) from producer to consumer. In a grid that utilizes diverse and smaller-scale renewable systems, energy must be understood to flow omnidirectionally. As Sørensen and Kahn point out (chaps. 6 and 8), the grid itself can serve as an equalizer of the ebbs and peaks of varying production. As the number of small producers (or contributors) to the grid grows, it is essential to understand the impacts. Sørensen argues that between 10 and 20 percent of a grid's total production can be produced from dispersed and variable sources before the need for storage becomes critical.

Dependence on the Marketplace to Determine the Cost Effectiveness of Various Production Alternatives

The problem of determining which sources and technologies to use to meet demand is a complicated management problem. As the proportion of dispersed to central generation grows, the problem will grow. The question that must be addressed is, "What technology can meet demand most cost-effectively?" As Huettner points out (chap. 9), the question may best be answered by using the marketplace as an analogy. In an emerging new role, the utility would become a broker between buyers and sellers of electricity. With smaller stations, the financial risks and costs of overcapacity to producers and consumers are reduced. Thus it may be easier for utilities to find new capacity in small-scale units. Tax incentives such as energy credits and accelerated depreciation allowances in combination with enabling legislation (see chap. 10) can encourage small-scale production. Huettner discusses this potential extensively in chapter 9. As the cost of capital rises for central capacity generation, and debt threatens the viability of utilities, they may well be increasingly interested in being relieved of their responsibilities and burdens for generation.

A Greater Degree of Institutional Separation between Production and Distribution

Increasing dependence on markets for production necessarily involves independent producers competing to sell as well as potential consumers competing to buy. The utility as broker and distributor would be less and less of a force in production. This approach requires redefinition of the traditional electric utility, but it is not without precedent. In the communications industry a "utility" is a public service industry responsible for distribution of a commodity (information). Such utilities are barred from controlling access to the network by producers or consumers except for the imposition of quality standards to protect the network itself. In the electric utility industry, utilities have historically been licensed and protected not only as the distributors but as the sole producers of the commodity they distribute.

In a marketplace, producers of an inefficient or overly expensive alternative are penalized with lack of sales. Electric utilities, on the other hand, have been rewarded with planning expensive overcapacity by rate increases based on guaranteed return on investment. This problem would be ameliorated by removing the rights and responsibilities from utilities (as Huettner discusses in chap. 11).

The Development of an Independent Service Industry to Manage Decentralized Production

Lindsley points out in chapter 13 that the primary problem with decentralized production is the reliability of individual plants owned by individuals and institutions without expertise in their operation and without capital to maintain parts and service. Service is a labor-intensive operation, and dispersed generation can work only with efficient and well-equipped regional service and management industries. A considerable degree of standardization (not now evident) in equipment design would be required for an economically viable industry over the long haul.

Some Expansion of Both Centralized and Decentralized Storage

Reliability will always be an issue as the percentage of variable production increases (i.e., with the growth of the photovoltaic industry). The need for storage and its characteristics are discussed by Robert Morris in chapter 5. But decisions about storage (how much and what kind) can be made only on the basis of market conditions, the effectiveness of load management, and the character of supply.

To a large degree, decentralized storage can be thought of as a form of load and end-use management. Wind may be better suited as a tool for pumping water and pressurizing air to displace electric generation because water and air are storage mediums. In such instances, the windmill serves as a conservation rather than a production system and variability is not a problem in such uses. Central storage of energy as hydrogen or pumped water are also options, but storage should always be considered a last resort in grid management because of its high costs.

The kinds of changes discussed in this book for our electric utility infrastructure will not come overnight, nor is it likely that the changes will be as clear-cut as suggested by this brief characterization. The ten characteristics discussed above are presented as directions in which change could take place. Many of the ideas, radical when this book was conceived, are now at least partially accepted in the industry. Small-scale, decentralized production is, under the federal and state guidelines discussed by Thompson in chapter 10, now growing rapidly, although its contribution is still relatively small.

NOTE

1. Howard Odum, "Energy, Ecology, and Economics," *Ambio*, vol. 2, no. 6, pp. 220–27 (Swedish Academy of Science, 1973).

AMORY B. LOVINS

CHAPTER 2 | *Technology Is the Answer! (But What Was the Question?)*

Amory Lovins presents a critical review of the electrical generation industry and of the concepts that currently guide decision-making in it. His chapter, written in 1978, is a personal observation in the sense that Lovins does not include in it all the details and documentation that he has presented to support his observations elsewhere. He concentrates on casting a shadow of doubt over the dominant assumptions underlying policy development in the power production agencies, private and public. Regardless of whether we agree with Lovins's proposals, the article is an excellent introduction to an alternative view of the inadequacies of the present electrical generation system. Lovins's years of research and analysis, carefully documented in other papers, show themselves clearly in this plea for a saner and more useful electrical generation system.

The energy problem is both important in its own right and useful as an integrating principle for examining a wide range of related resource and social problems. I am not concerned here primarily with the technical features of various energy technologies, however seductive, but rather with their appropriateness, their fitness for specific tasks; and having studied them from various perspectives, I shall conclude that what I call "hard" energy technologies are, in Marvin Goldberger's memorable phrase, spherically senseless—that is, they make no sense no matter how you look at them.

While not under the illusion that facts are separable from values, I attempt in this critique to separate my personal preferences from my analytic assumptions, and to rely not on modes of discourse that might be viewed as overtly ideological but rather on classical arguments of economic and engineering efficiency and of orthodox political economy (arguments which are only tacitly ideological). The residual disagreements to which the results may give rise are

Copyright © 1983 by Amory B. Lovins. All rights reserved.

in general due to transscientific differences that cannot be resolved on technical merits. Such disputes, often masquerading in technical guise, dominate the energy debate, and failure to recognize them for what they are causes no end of confusion. Nor do I wish to imply, by emphasizing economic arguments, that I consider them dispositive or even especially important. A dispassionate analysis of how we actually make major public policy decisions about energy would reveal that we decide on grounds of political expedience and then juggle the subsidies to make the economics come out to justify what we just did. Nonetheless it is a sound tactic to use one's opponents' data and criteria so that those who prefer to count only what is readily countable (private internal costs) will find their accustomed analytic methods, if not their conclusions, given gratifying emphasis here.

Though this is essentially a critique of traditional, currently dominant approaches to the energy problem, I have taken the liberty of adding at the end a very brief sketch of a "soft" energy path that appears to be more justifiable and more likely to succeed.

HARD ENERGY PATHS

The traditional energy policy of Strength through Exhaustion—converting increasingly elusive fossil and nuclear fuels into rapidly growing amounts of premium energy forms (fluid fuels and, especially, electricity) in ever more complex and centralized plants—is inappropriate because, most fundamentally, the tasks for which it was to be appropriate were left undefined. The energy problem was thought to be how to expand secure, affordable, and (preferably) domestic energy supplies to meet extrapolated homogeneous demands. Demand was treated as an aggregate figure (so many quads in 1984) without regard to the most effective type or scale of energy for each end-use task. The result, extravagantly, was a highest common denominator, an array of costly and elaborate trip-hammers capable of cracking any conceivable nut.

More specifically, the chain of argument underlying this approach runs something like this:

1. To meet our social goals (however unspecified or platitudinous) we need
2. rapid undifferentiated economic growth, which requires
3. more or less correspondingly rapid growth in primary energy use, so
4. we rapidly run out of (that is, encounter increasing economic, geopolitical, or ultimately geological difficulties in obtaining) oil and gas, so
5. we must switch to the more abundant solid fuels (coal and uranium), but
6. direct use of coal is not generally feasible or convenient, so
7. we must burn the solid fuels in power stations (and perhaps, ultimately, in coal-synthetics plants), and because of
8. economies of scale

TECHNOLOGY IS THE ANSWER!

9. we need the power stations to be big, so
10. the only question is which kind of big electric plant to build, and the canonical answer is
11. nuclear (and perhaps coal-fired)—built rapidly and profusely.

Embarrassing questions, raised with increasing force and frequency, are perforating this hermetic argument. For example:

2. With respect to economic growth, what is growing? What has it to do with welfare? Is its net marginal utility positive? How do we know? How long will it stay that way? If our welfare derives from material rather than from cultural or spiritual things (a bad approximation), should we not try to maintain the maximum stock of physical artifacts with the minimum throughput of resources and effort, and if so, is not most of the GNP something we should try to minimize rather than maximize?

3. What is the link between (2) and (3), since we now know that by practical, economically attractive, and purely technical measures we can double by about the turn of the century, roughly redouble by 2025, and further increase thereafter the amount of work wrung from each unit of delivered energy—i.e., that within very broad limits energy and economy can be decoupled?

7. How can we afford the power stations on a truly large scale (large enough to substitute nationally for oil and gas), since central-electric systems are two orders of magnitude more capital-intensive than historic direct-fuel systems and have a very unfavorable cash flow to boot? How can we realistically expect electricity to penetrate the markets accounting for most of our delivered energy needs—heat (58 percent) and portable liquid fuels for vehicles (34 percent)—in view of (a) its cost (new electricity costs more than a hundred dollars a barrel on a heat-equivalent basis, or several times the present OPEC oil price) and (b) the formidable rate and magnitude problems of the complex, slow-to-deploy electric systems? (Supplying with nuclear power a quarter of the lowest government projections of United States energy needs in 2000—assuming that each unit of electricity replaces two units of fossil fuel throughout the economy—would require us to order a 1,000-MW station, starting now, every 4.7 days. This would require more investment than we now put into all industry. Further, since nuclear power, even in principle, can readily displace only baseload electricity—a small fraction of all our energy, and oil, uses—it cannot do much for oil dependence: replacing overnight with nuclear power every oil-fired power station, both thermal and gas-turbine, throughout member countries of OECD [Organization for Economic Cooperation and Development] would reduce 1975 OECD oil consumption by only 12 percent and would reduce the fraction of consumption that is imported from 65 percent to 60 percent.)

11. Is this result politically, geopolitically, environmentally, and economically acceptable? To whom? Is that enough consensus to proceed on (cf. Vietnam)?

But it is not my purpose here to canvass these interesting arguments, or the more technical ones concerned with (5) and (6). Rather, I shall concentrate on (8–10), considering

8. countervailing diseconomies of scale,
9. structural problems of central electrification, and
10. our ability to make political decisions of this kind.

Diseconomies of Large Scale

Real economies of scale frequently occur in construction. These are often said to follow a scaling law such that increasing unit size by a factor X increases cost by a factor only X^n, where n is of order 2/3. (In practice the observed economies are often lower, with n approaching unity in some cases.) But there are also countervailing diseconomies of large scale that have seldom been properly taken into account, often because they are outside the system boundary represented by X. They include:

DISTRIBUTION COSTS. If we make an energy device, such as a power station, refinery, or gas plant, bigger and more centralized, we must pay for a bigger distribution system to spread the energy out again to dispersed users. In the United States we have gotten to the point, with electricity in 1972 and gas in 1977, where an average residential customer was paying about 30 cents of each utility-bill dollar to buy energy and the other 70 cents to get it delivered. That is a diseconomy of centralization. (Utilities tend not to notice it because they take incremental distribution costs as given and seek the cheapest source of bulk electricity to feed into the "preexisting" grid—even though, in practice, minimizing busbar cost may maximize delivered price.)

DISTRIBUTION LOSSES. These are generally of the order of a tenth of throughput, but they are pervasive.

LOSS OF OPPORTUNITIES FOR MASS PRODUCTION. If we could make power stations the way we make cars, they would cost at least an order of magnitude less than they do, but we can't because they're too big.

LOSS OF OPPORTUNITIES FOR INTEGRATION. Total-energy systems (for example, cogeneration, combined heat and power stations) and integration with, for example, food and water systems can save a great deal of money but are generally impracticable at the scale of modern central energy facilities.

UNRELIABILITY. Big plants (notably power stations) tend to be less reliable than smaller ones, for excellent technical reasons that are not likely to go away. (For example, a 500-MW boiler has about ten times as many miles of tubing as a 50-MW boiler, so it will fail more often unless quality control improves tenfold; physically larger turbines have larger blade-root stress and hence require more exotic alloys more likely to have unexpected properties.)

LARGER RESERVE-MARGIN REQUIREMENTS. Failure in a 1,000-MW power station is embarrassing, like having an elephant die in the drawing room, and requires a second elephant standing by to haul the carcass away (1,000 MW of backup capacity). This is expensive. More numerous smaller stations would be unlikely to fail all at the same time and hence would need less reserve margin: in practice, changing unit size from 1,000 MW to a few hundred MW would provide the same level and reliability of service with about a third less new capacity, and 10-MW units at the substation could save over 60 percent at the margin. With realistic assumptions about reserve requirements *and* unreliability as a function of size, reserve requirements can change very roughly as the square of changes in unit size. With each big new station costing several billion dollars, the incentives to save a third or more of that capacity are high.

HIGHER INDIRECT COSTS. There is some empirical evidence that installing a kW(e) in a small and supposedly uneconomic thermal power station can actually cost less capital than installing a kW(e) in a very large station. This is probably because the small station is so much *faster* to build that it greatly reduces exposure to interest payments, cost escalation, changes in regulatory requirements during construction, and the risk of premature completion.

LOSS OF DIVERSITY. Big units make it possible to make truly large mistakes at high social and economic risks. Long lead times (tempting one to compress development and scale-up schedules) and technical adventurousness compound the risk of large-scale technical failure. While small systems adapted to particular niches can mimic the strategy of ecosystem development, adapting and hybridizing in constant coevolution with a broad front of technical and social change, large systems tend to evolve in a more linear fashion, like single specialized species (dinosaurs?), with less genotypic diversity and greater phenotypic fragility. Adaptation is further constrained by the accretion of a costly and inflexible infrastructure.

INABILITY TO DISTINGUISH AMONG USERS. People who use electricity for heating water and would not even know if it went off for a few hours must pay the high premium for the reliability required by elevators, subways, and hospital operating theaters. For the former group this is a large diseconomy.

TECHNICAL THRUST TOWARD INFLEXIBLE DESIGN CRITERIA. For example, it is not obvious that a future electric grid operating on dispersed renewable sources (hydro, microhydro, wind, photovoltaics, solar heat engines) will need or be able to justify high standards of frequency stability and phase coherence: with little rotating machinery, it may be worth making a cheaper, sloppier grid. (Frequency stability is, very roughly, five times as good in the United States as in Western Europe, and five times as good there as in Eastern Europe.) Customers who want to use their power supply as a clock could instead use local oscillators (now very cheap) or radio markers like WWV (a National Bureau of Standards shortwave broadcast station providing continuous, precise time and frequency reference signals). In contrast, some new steam turbines now being ordered are so inflexible that they blow up if their operating frequency deviates by a rather small fraction of one percent. Installing such devices locks us, for many decades hence, into very costly and perhaps unnecessarily stringent operating criteria.

VULNERABILITY. Big units increase the tendency of central electric systems (and also nonelectric analogs) to be vulnerable to disruption, whether by accident or malice. In a centralized grid a few people can turn off most or all of a country, whereas dispersed sources, while benefiting from user diversity on the grid, would not be dragged down if one or two sources failed: in many instances local users could still continue to use local sources decoupled from the grid. Central electric systems can be designed for high technical reliability in the face of calculable failures, and are (at a high marginal cost), but they tend not to be resilient in the face of incalculable (but numerous and important) surprises.

INCREASED LOCAL SOCIAL AND ENVIRONMENTAL STRESS. This makes site licensing more difficult, so utilities seek to maximize installed capacity per site, so the plant is a worse neighbor than it would otherwise have been, so the political reaction raises transaction costs for the next site, and so on exponentially. We are well into this loop. (It can be argued that more dispersed sources are harder to submit to environmental controls. European experience contradicts this. Smaller units can often use inherently cleaner technologies, be integrated into total-energy systems that minimize fuel burned per unit function delivered, and, since they are sited amongst their users, must be and are built and run properly—whereas central, remotely sited units often have the political clout to alter or ignore environmental controls.)

HIGHER COMPLEXITY. Hence longer downtime, more difficult repairs, higher training and equipment costs for maintenance, higher carrying charges on costly spare parts made in small production runs, etc. Management also may become more complex, with high fixed charges encouraging haste and corner-cutting. (Ten percent annual interest on a billion dollars is over three dollars per second.) Technical problems, interacting with environmental, social, economic, and political problems, are often unforeseen, especially in a world of lags, nonlinearities, threshold effects, irreversibilities, electoral cycles, and other peculiarities unlikely to be foreseen by plant designers and operators. These surprises tend to promote the already dangerous increase in the likelihood and consequences of mistakes.

Another category of structural problems associated with centralization, especially (for illustration) in electrical systems, often manifests itself in higher costs:

CENTRALIZATION AND AUTARCHY. Allocating enormous amounts of scarce resources to such a demanding enterprise in the face of competing claims, especially when the market is unwilling to do so, requires a strong central authority —an Energy Security Corporation (to evade market forces) and an Energy Mobilization Board (to evade democratic forces). Big, complex energy systems require big, complex bureaucracies to run them and to say who can have how much energy at what price. The macroeconomic side-effects of extraordinary capital intensity (for example, inflation, unemployment, high interest rates) elicit further central management, chiefly by distortion (for example, further subsidies), taking ever more bizarre forms in an effort to protect a sector too big to allow to crash.

ENCOURAGEMENT OF OLIGOPOLY. Small business can't make big machines.

IRRELEVANCE TO THE NEEDS OF MOST OF THE PEOPLE IN THE WORLD. Big, elaborate electrified systems are probably the least sound energy investment for developing countries, though elites often desire them for prestige—largely because of the bad example set by the industrialized countries. In my view, for example, the greatest contribution of United States nuclear reactors to the global problem of nuclear bomb proliferation is undoubtedly the example they set, whereby all countries feel entitled to similar reactors and the materials associated with them—an entitlement affirmed by every president since Eisenhower. It is unconscionable for this country to rely on an energy source we decry as proliferative elsewhere—albeit hypocritically, in view of our own reliance on nuclear bombs—and we cannot expect that if we, with our fossil fuels, money, and skills, continue with nuclear power, we will see restraint from other countries lacking these advantages. But that is another paper. It is noteworthy that most electrical equipment vendors are so oriented toward large-scale OECD markets that they are unable or unwilling to supply small units for developing countries, which must therefore buy large units and bear the consequences—such as the blackout in Thailand on 18 March 1978, in which 33 million people lost power for up to 9 hours in Bangkok and 56 other provinces owing to the failure of one 1,300-MW(e) plant in a 2,500-MW(e) grid. In contrast, cooperative marketing would be a natural and mutually beneficial trend for smaller systems, especially those suited to rural development (photovoltaics, wind power, solar heat engines, etc.).

ALLOCATION OF COSTS AND BENEFITS TO DIFFERENT GROUPS OF PEOPLE AT OPPOSITE ENDS OF THE DISTRIBUTION SYSTEM. This produces strong interregional conflict.

CENTRIFUGAL POLITICS. Central siting and regulatory authority versus local autonomy is already a potent constraint on energy expansion and is threatening severe strains in the federal (and state/local) system here and abroad.

ENGINEERING INEFFICIENCY. Electricity often cannot in practice be used to its theoretical thermodynamic advantage. Owing largely to proposed electrification, more than half the projected primary energy growth in most industrial countries would be lost in conversion and distribution before it ever got to final users.

IMPLICATIONS FOR TECHNOLOGISTS. With technologies requiring many years' effort in big, anonymous research teams, personal responsibility and initiative are diminished and may even slip through the cracks altogether. Powerful promotional constituencies develop and take on a life of their own. Further, as Freeman Dyson points out, big technologies are less fun to do and too big to play with, so technologists may not be as innovative as with smaller things that lend themselves to tinkering. Indeed, because soft energy technologies are so accessible that one person, even without much technical training, can make an important contribution to them, they can profit fully from human diversity. There is, as far as we know, nothing in the universe as powerful as four billion minds wrapping around a problem. That is the source of the extraordinarily rapid progress we are now seeing in soft technologies.

Finally, and blending with earlier categories, some special problems of political economy beset centralized (especially electric) systems and tend to raise their economic costs.

INCOMPREHENSIBILITY. In what Rob Socolow (in *Patient Earth* [New York: Holt, Rinehart, and Winston], 1971) has called "consumer humiliation," users are compelled to depend on systems they cannot understand, modify, repair, or control. They are then told that it is desirable for them to stop worrying and leave it all to the experts; but occasional reminders that experts are fallible leave a sense of unease and, for some, alienation. Some people like to understand their own systems, and most, I think, suspect that systems allegedly too complex for ordinary people to make up their own minds about are systems which a democracy ought not to have, for they remove any chance of accountability and substitute for the democratic process a sort of . . .

ELITIST TECHNOCRACY. "We the experts" replaces "we the people"—gratifying for the experts but likely to lead to a loss of legitimacy, a dangerous and hard-to-reverse trend that rubs off elsewhere.

INEQUITABLE ACCESS. Remote siting, which unloads social costs on politically weak agrarians (Navajos, Wyoming ranchers, Montanans, Alaskans) to provide energy to politically strong slurbians (Los Angelanos), is bad enough; worse are technologies so arcane, complex, large, or costly that only wealthy people or large corporations stand much chance of benefiting from them. (This is especially a problem in developing countries: if a new power station is built in India, for example, roughly 80 percent of its output tends to go to urban industry, 10 percent to rich urban households, and 10 percent to villages—and in the end 1 percent might end up helping the poor people for whom it was ostensibly intended. Electricity from any source is costly and hence unlikely to reach those people—some two billion at least—who are outside the market system and have neither an electric outlet nor anything to plug into it.)

HIGH INERTIA. A utility investing 2–6 billion dollars in a reactor wants to be certain it can get an operating license in 1990 and operate until 2030 without onerous changes in conditions—and wants that certainty in a society where we keep changing our minds and changing our values, and we even throw the rascals out every couple of years. We cannot give such certainty, and investors who want low risk should seek low-impact, understandable, short-lead-time—in short, as Clark Bullard asserts, *low-inertia* technologies that can adapt to rapid changes in values (and in technical conditions).

HOMOGENIZATION. The high costs of manufacturing billion-dollar units are a strong incentive for standardization (whereas, say, solar collectors can benefit from mass production and still be adapted to local conditions). High capital intensity is an incentive to adapt people's energy needs and patterns of use to the convenience of the technology, not the reverse. The high fixed costs and inflexibility of electrical infrastructure (or, say, of gas pipelines) tend to lock us into particular settlement patterns, end-use technologies, and habits—or risk incurring high costs for modification.

PARAMILITARIZATION. Not only the inherent vulnerability of centralized systems (especially those that rely on a form of energy not conveniently storable in bulk) but also their explicit links with military applications can encourage social controls that threaten to abridge traditional civil liberties. While nuclear power is so far the best example, the same may well be true of, say, liquefied natural gas terminals, electric load dispatching centers, major fossil-fuel facilities, and other potential leverage points attractive for saboteurs. Severe conflicts between the aims of labor unions and security-conscious managers are also likely (and already emerging in several areas).

These disadvantages of large-scale energy systems—some obvious, some subtle, some readily quantifiable, some fuzzy, and most interactive—do not mean everything should be small. But they do mean that, in order to minimize costs (including social costs), energy systems should be scaled to be appropriate to their task. This is already a familiar concept in transportation, where the diversity of settlement patterns and life-styles in a pluralistic society yields a spectrum of transport densities needed. Mass transit (often electrified) is most appropriate for high-density intra- and inter-city commuting; variable-route public transport and private modes are most appropriate in medium-density suburbs; and private vehicles and a sprinkling of other modes are best suited to the countryside.

Mode should match need, rather than, for example, trying to bring a dense subway network to the countryside or smother the inner city with cars. Just so, a rational energy system should use large existing hydroelectric dams for smelters and small solar collectors for single houses, rather than trying to run smelters with little wind machines and heat houses with fast breeders—a symmetrically nonsensical mismatch. This matching of the spectrum of energy and transport density supplied to that required is the essence of appropriate-technology thinking—not the predetermined predominance of a particular scale, small or large. That is why conservationists are consistent in seeking both generally smaller energy systems and more urban mass transit, and why their adversaries, with equal consistency, tend to seek both gigantic energy systems and the universal primacy of the private car. Appropriateness needs a shift toward larger average scale in transport, smaller in energy (where scale is irrationally large by many orders of magnitude for all but a tiny handful of uses); but these shifts are a consequence of a general principle of appropriateness to the task, not the other way around. Each energy system or transport system should do what it does best, not try to do something else or be a panacea. In this sense it is the hard path, with its bias toward homogeneity and large scale, that is ideologically rigid, and the soft path, with its emphasis on fitness for the task at hand, that reflects a flexible, pluralistic social fabric.

It would be wrong to suppose that the structural problems of hard energy technologies arise only from their inappropriate scale. At least as important is their inappropriate bias toward high-quality (hence costly and hard-to-make) forms of energy, notably electricity, in a world where needs for which those forms are appropriate have long since been saturated, and where the dominant need (heat and portable liquid fuels) can be met more cheaply by simpler and more direct means. Indeed, I view the thermodynamic matching of energy supply to the spectrum of end-use needs as more important conceptually than scale matching. The saturation (by a factor of two in the United States today, probably over four in the long run) of electricity-specific end-use needs means that more electricity from any source is not a rational response to the energy problem we have: it is too slow and too costly, to say nothing of its unpleasant side effects. Hence arguing about which kind of power station to build is akin to debating the best buy in champagne when all one wants is a drink of water. The same appears to be true throughout the industrial world, and probably in many developing countries too.

REDEFINING THE ENERGY PROBLEM

An appropriate-technology approach to the energy problem begins by asking what the energy system is to be appropriate *for*: that is, what tasks are we trying to do with the energy, and how can we do each task with an elegant frugality of energy (and other resources) supplied in the most effective way for each task? These unfamiliarly simple questions lead naturally to a different evolutionary path for the energy system—one I have called a soft energy path. It has three technical elements:

- greatly increasing end-use efficiency (wringing several times as much work from each unit of delivered energy; I assume only technical improvements to this end, not any significant changes in life-styles, settlement patterns, patterns of social or economic organization, composition or growth of GNP, etc.);
- rapidly introducing soft energy technologies (diverse renewable sources that are relatively easy for the user to understand and that supply energy in the scale and quality appropriate to each end-use need);
- at the same time using fossil fuels briefly and sparingly, in clean transitional technologies so designed that soft technologies can readily be plugged into their infrastructure as they come along.

My analysis suggests that in the United States—and, indeed, in over a dozen other countries spanning a wide range of conditions—reliable and convenient soft technologies that are already in or entering commercial service, that is, that are here already and do not require further research and development, are more than enough to meet virtually all our asymptotic energy needs (though it will still take a transition of about fifty years to get them all in place, because the energy system is so sluggish that it takes that long to do anything). I include in this analysis the present art (if one shops around carefully) in passive heating and cooling, active solar heating with enough storage to need no backup, solar process heat for industry, methods of converting farm and forestry residues to liquid fuels (which are then enough to run a very efficient transport sector without using special "energy crops"), present hydroelectric capacity, microhydroelectricity, and a modest amount of wind (for electricity, water pumping, heat pumping, and hydraulic drive). I do not include soft technologies for solar electricity—such as cheap photovoltaics (which will be here before we know what to do with them, if indeed they aren't here already) or solar ponds and heat engines (which are already commercial and look cheaper than any conventional baseload source)—because we don't really need them, though they will make life much easier. Many of the designs for the soft technologies I do assume show subtle and sensitive interactions with the details of end use: for example, a very energy-efficient house (the first priority no matter how it is to be heated) requires a very different, and much cheaper, kind of solar heating system than a normally leaky house. Few analysts are aware of these interactions between demand and supply design.

In analyses published elsewhere I have argued that compared with a hard energy path, a soft path is

- cheaper (calculated conservatively from empirical cost and performance data; there is a large advantage in capital cost and in cash flow, and an even larger advantage in delivered energy price);
- faster (yielding a greater and quicker return in energy, money, and jobs per dollar invested);
- environmentally much more benign (including hedging our bets on the CO_2 problem by getting out of the fossil-fuel-burning business as quickly as possible);
- surer to work (spreading the risk among dozens of technologies already known to work, not relying on a few that may or may not work);
- more compatible with modern concepts of development;
- a strong political lever for nonproliferation, rather than the opposite; and
- politically much more attractive, offering simultaneous advantages to most constituencies and building on an existing consensus.

The "faster" contention deserves a little amplification. It arises because the tools of a soft path, including soft technologies, are so simple that their lead times are one to three orders of magnitude shorter per unit than those of hard technologies; are so accessible that they sell to a market orders of magnitude larger (households versus utilities, for example); and are so diverse—with dozens of relatively slowly growing components, each retarded by constraints relatively independent of the others'—that by sheer strength of numbers they can add up to very rapid total growth. This is not at all the same as having a few monolithic technologies subject to generic rate constraints; but classical market penetration models cannot distinguish the cases.

As a thought experiment, consider two kinds of wind machines, both now in or ready for commercial production. The first puts out 2 MW(e) and can be built by a utility in six months to a year. The second puts out nearly 20 kW(e) and is so designed that a couple of Midwestern farmers can bring it home in the pickup, then put it up in fifteen hours with a screwdriver and two crescent wrenches, plug it into the fuse box, and begin operation. Clearly the second makes up for its smaller unit size by short deployment time and potential for mass production; but, interestingly, it may also make it up by its wider market: there are several thousand times more farms than utilities in the Great Plains.

The appropriate market penetration model would thus be one designed for things less like basic oxygen furnaces than like citizen's-band radio (13 million Americans and growing by half a million a month as of early 1977); things less like jumbo jets than like snowmobiles and digital watches.

There is good anecdotal evidence that people can do many relatively simple things for themselves very quickly if they have incentive and opportunity. For example, during the years 1974 to 1976 some 40 percent of all Vermont households retrofitted wood-burning iron stoves (not a trivial operation), entirely on their own initiative, because they had the chance to do so and the oil price provided the incentive. The process was more akin to spontaneous technical diffusion than to the "technology delivery" that the Department of Energy (DOE) worries about. During 1971 to 1976 the fraction of American households trying to grow some portion of their own food (even tomatoes in the windowbox) reportedly rose from about 20 percent to 50 percent—again without official programs. In other cultures, where admittedly I would not like to live, there are even more spectacular examples—such as the 4–4.5 million biogas plants built in China since 1972. From my sense of human potential and of actual grass-roots energy activities in this country, I judge that a soft energy path could implement itself largely through existing market and political processes if it were allowed to show its natural economic and social advantages. This implies

- clearing away a long, messy list of institutional barriers to energy efficiency and soft technologies (3,000 obsolete building codes, obsolete mortgage regulations, inequitable access to capital, restrictive utility practices, split incentives between builders and buyers or landlords and tenants, bad information, professional fee structures that encourage inefficiency, etc.);
- desubsidizing conventional fuels and power (cumulative subsidies are several hundred billion dollars, and the current rate is tens of billions per year transferred from our energy bill to our tax bill);
- moving gradually and fairly, as I think we know how, toward charging ourselves for depletable fuels what it costs us to replace them in the long run.

Though it will not be easy to do any of these things, I think it will be easier than not doing them, and, if properly handled, can have great political appeal.

The soft and hard energy paths are distinguished not just by how much energy we use, nor only by choices of energy equipment, nor even by what they assume the energy problem is (more energy of any kind versus the right kind and amount to do each task in the cheapest way), but also by their very different *political* implications. The hard path is one whose polity is dominated by such problems as centralization, autarchy, vulnerability, and technocracy; the soft path has a different set of more tractable problems (chiefly those of pluralism), and we must decide which kinds of problems we prefer. A hybrid hard-and-soft path is a category mistake, at least within these definitions, because the political conditions that define the two paths cannot logically coexist in the same society. Hard and soft *technologies*, however, are not technically incompatible: indeed, in a soft path they would coexist during the fifty years or so of gradual transition from the present mix (nearly all hard) to the asymptotic mix (practically all soft). This evolution would have to take place within a social and political context. As a result the two broad evolutionary patterns, predicated on different perceptions of the nature of the energy problem, would become mutually exclusive:

- They are culturally incompatible: each world makes the other world harder to imagine. Where we are now illustrates this well: our efforts of the past few decades have given us a cadre of people who simply cannot imagine any approach to the energy problem other than what they've been doing.
- They are institutionally antagonistic, each requiring laws, organizations, and policy actions that inhibit the other, just as our rigidities and institutional barriers are now locking us into where we've been rather than where we may want to go.
- They compete for resources: every bit of work and skill, every dollar, every barrel of oil, every year that we commit to the very demanding hard technologies postpones taking on efficiency improvements and soft technologies. There is thus a risk that if other commitments push the era of substantial soft-technology contributions too far into the future, the fossil-fuel bridge may be burned first.

This argument suggests that we should, with due deliberate speed, be choosing one or the other of these broad paths before one has foreclosed the other (or before proliferation has foreclosed both), using thriftily the cheap fossil fuels—and the cheap money made from them—to finance a transition straight to our ultimate energy-income sources, because we won't have another chance to get there.

BROADER REFLECTIONS

I have not tried to carry the definition of soft energy technologies into wider applications, though it clearly has some affinities with appropriate-technology approaches in other fields. Nor have I tried to solve all the social and political problems that have exercised philosophers for the past few millennia. My aim was rather more modest than that: to find design criteria for the energy system that use existing institutions and incentives to solve pragmatic problems without causing too many more. I think the resulting approach is likely to make many other kinds of problems in our society easier to solve, but it will not automatically dispose of them. Further, a soft energy path as I have analyzed and advocated it is not an instrument of social change but rather a way of *avoiding* social change—of the nasty kinds we otherwise get with a hard path—while preserving the status quo if that is what we want. It offers an opportunity for, and is compatible with, a more Jeffersonian society, but in no sense entails one.

It is likely that the future I have assumed for my analysis is macroeconomically inconsistent, in the sense that it assumes (for purposes of argument) continued rapid GNP growth of a conventional kind, and presumably continued growth in exports, at a time when we would no longer be needing foreign exchange to buy oil and gas. Greater self-reliance in energy has important implications for the rest of the economy which nobody has yet thought through. Further, it seems to me that as we move from a world in which we bought raw materials at competitively depressed prices and sold manufactures at monopoly rents to a world that is increasingly the other way around, more such trade will merely dig us into a deeper hole. While there are many respects in which the classical economic-growth-and-free-trade model is no longer satisfactory—regional disparities high on the list—perhaps the most interesting is its failure to account for the increasing advantages of internal as against external transactions. If I can buy a widget for $1.50 made locally or for $1.00 from a faraway industrial center, it may be to my advantage to buy it locally and so keep the money *chez nous* and get the local multiplier, rather than pay for someone else's middlemen and transaction costs. Fraser Darling, in *West Highland Survey: An Essay in Human Ecology* (Oxford: Oxford University Press, 1955), describes a sea loch with identical conditions on both sides save that it had a road on one side and a dirt track on the other. The former side had a low standard of husbandry, declining fertility, little fishing; the people ate little of their own produce but sold it to Glasgow, seventy miles away, in order to buy tinned porridge and packaged bread. The land was becoming run-down because, while the road had tied the farms into the commercial web of the south and east (hence into its transaction costs), it had not so reorganized the habitat as to enable the farmers to pay for this connection by sustainably exporting more. On the roadless side of the loch importation was possible where advantageous, but it was not obligatory. The land was in good shape, the people ate largely their own oatcakes and fish and dairy products (and hence were healthier), and the whole enterprise was sustainable and flourishing. Perhaps that loch offers a metaphor, albeit an oversimplified one, for how the logic of national balance of payments can reproduce itself on a regional, state, local, and even household level.

The last comment I should like to draw out of the soft-versus-hard energy example bears on the way we think about technologies. A soft energy path would be in no sense an antitechnology program; it would involve much exciting technical challenge, but of a different and, to some, an unfamiliar kind, making things sophisticated in their simplicity rather than in their complexity. This raises an important point which I think is largely missed by those who talk of "high" versus "low" technology, or "sophisticated" versus "simple" technology, or "advanced" versus "intermediate" technology—where the former term in each pair is used as a diffuse honorific and the latter used somehow to imply backwardness. Perhaps it is the definer who is backward. When my father was in engineering school, it was always drummed into the students that any fool can make something complicated but it takes genius to simplify. Since World War II we seem to have lost much of that perception. (A Soviet friend recently told me his country has the same problem—they too have their equivalent of Bay Area Rapid Transit (BART)—and he attributes it, as I do, largely to the flow of military and aerospace engineers into civilian sectors.) We have now got to the point where the first design for the Boeing-Vertol subway cars in Boston reportedly had 1,300 parts—eventually pared down to 300, but evidently the designers had become so "sophisticated" that they couldn't design a door any more. A recent conversation with an aluminum dealer revealed to me that the floors of the BART cars in San Francisco are made of aluminum plate because that is what the airframe designers were used to (though steel would have worked much better); then fancy extrusions had to be added on the edges because aluminum isn't stiff enough; then the aluminum had to be anodized, then painted, and then carpeted. My informant wisely concluded that anyone who could design a vehicle that way was too risky to work with, so he declined to tender to supply the flooring. He seemed pleased in retrospect with his decision.

These anecdotes reflect, I think, a split in the technical community that may well be the most serious obstacle to a soft energy path and, more broadly, to appropriate technology. It is a split between those who can appreciate elegantly simple design solutions and those who can't. The NASA engineers who designed the ill-fated Sandusky wind machine, for example, reportedly spent several tens of thousands of dollars on a computerized system of electronic instruments to shut down the machine if it began to vibrate too much. The Danish engineers who built the Gedser mill many years ago had solved the same problem with a much higher order of engineering: they had in the tower a wide saucer with a big steel ball in it, and if the tower vibrated too much, the ball would slop out of the saucer and fall down, and a string attached to it would then pull a switch. That kind of simplicity is virtually unheard-of in official energy programs. Maybe it's *too* sophisticated.

As an experiment, when speaking to high-technology audiences I occasionally describe two simple energy devices now operating in New England:

- a solar pond consisting of an insulated hole in the ground, 1 m deep by 5 m across. In the bottom sits a layer of coal as a black absorber. Above it is a thick layer of a strong $CaCl_2$ brine containing a few loops of pipe as a heat exchanger. On top is about 15 cm of fresh water. It sits there in the open. The brine sets up a self-stabilizing gradient of refractive index, temperature, salinity, and density that acts as a fisheye lens, converging the whole hemisphere of sky onto the coal. Out of the heat exchanger comes over two kW of continuous heat, above 90°C.
- a passive solar greenhouse consisting of three insulated walls and a south wall that is a slant roof made of three air-spaced layers of a fluorocarbon film that traps infrared. There is a convective rockbed in the floor. When I visited it on a bitterly cold February day, inside it was 30°C (86°F) and they were growing papayas.

When I ask high-technology audiences who among them is impressed and excited by these devices, typically half raise their hands (a tenth in electronuclear audiences a year ago, though half today—the climate is rapidly shifting). The other half, presumably, cannot get excited about these devices, which achieve complex effects with a great economy of means, because they are not big and electric, are not computer-designed, do not use exotic materials, and do not have brass knobs all over them.

Just one more example. When the Americans captured the German rocket factories at Peenemünde, their rocket experts reportedly tried to reproduce the German art by building V-1 buzz bombs; but as the American prototypes used up their fuel, their noses pitched up and they overshot the target, so it was necessary to add a baroque control system to correct this tendency. The designers apparently complained to the captured German engineer about this, remarking scornfully that he, from a nation supposedly so skilled in metallurgy, had also used such an inferior vane material for the flaps on the front of the pulse-jet engine that they started to disintegrate halfway through the flight. The German replied indignantly that he had taken pains to use precisely the alloy and heat treatment which would produce this effect, so that the power of the engine would diminish in phase with the fuel exhaustion and so make the control system unnecessary. That, I think, epitomizes (though hardly in the context of nonviolent technology) the kind of sophisticatedly simple engineering that our most specialized engineers could contribute to appropriate technologies if they were so minded. How to get them excited about doing so is not clear to me, though there are some encouraging signs that once a few bright people in a technical organization with a waning sense of purpose set up a soft-technology cell, others are likely to drift into their orbit. If this does not occur, it will be necessary to find narrow and unimaginative engineering tasks sufficient to occupy until retirement age* the engineers who don't want to seek imaginatively simple design solutions; and though there are doubtless a great many such tasks, not least in the terminal phases of nuclear power (a business with a very long future), it seems a waste of talent to rely on them exclusively when there are more worthwhile things to do.

Recycling the talent locked up in forlorn technical enterprises, and building in the technical community a coherent alternative vision of what can be done and why it is more interesting, seems to me one of the most difficult and urgent challenges of the transition to technologies that are spherically sensible. These technologies may be "high" in the sense that they call on extraordinary skills in design and even in manufacturing. But what I think matters most, especially in countries where such skills are readily available, is that the result should be a tool, not a machine. My pocket calculator, for example, is a highly sophisticated device in a technical sense; I don't know what goes on inside, and I couldn't

*This may be a long time, as there is no correlation between age and imagination: some of the best soft technologists I know are among the oldest, and some of the most narrow and rigid engineers I know are among the youngest, those brought up with the modern view that anything simple lacks technical sex appeal. This difference in training and temperament produces many interesting conflicts within the profession, as when younger engineers propose to base major decisions on computer simulations of systems whose phenomenology is poorly understood, while their elders, for whom engineering is an empirical art, wring their hands over the computer printout and say despairingly, "Yes, but we don't really *know* that." The all-too-common pattern lately has been for the former group to prevail and the latter to have to pick up the pieces.

make one. But from my point of view as a user, what matters is that I run it; it doesn't run me. It is, in that pragmatic sense, understandable and appropriate, not some arcane giant lurking over the horizon and run by a technological priesthood. As Fritz Schumacher said of cranks (when someone accused him of being one), it is simple, nonviolent, and causes revolutions. And, mindful of power failures, I retain as backup an intermediate-technology slide rule and a soft-technology brain.

Perhaps we shouldn't get too hung up on classifying the skills that go into making an artifact: brains take billions of years to design and are such a high technology that we don't understand them at all, but that doesn't mean we can't use them conveniently. What matters more may be the process that applies those skills to meeting human needs, as perceived by the final users. Is the process comprehensible, accountable, appropriate, sensible? I am reminded of the American woman living in India who called in a carpenter to repair a window frame, but he followed her sketch too literally and botched the job. When she remonstrated with him, asking why he hadn't just used his common sense, he drew himself up and replied with great dignity, "But common sense, Madam, is a gift of God; I have technical knowledge only." "Technical knowledge only": perhaps a good epitaph for a civilization. But I hope we have common sense, a grasp of fitness for the task, that will let us do better than that.

Since Amory Lovins wrote this in 1978, we have indeed done better. He calculates that between 1979 and 1982 the United States got more than one hundred times as much new energy from savings as from all expansions of energy supply; more new energy from renewable sources than from any or all of the nonrenewables; and more new electric generating capacity ordered from small hydroelectric plants and windpower than from coal or nuclear power plants or both, without even counting their cancellations.

DAVID MORRIS

CHAPTER 3 *The Pendulum Swings Again: A Century of Urban Electric Systems*

In expanding on the concepts discussed by Brown and Lovins, David Morris places today's electric industry in a historical context and observes that the decentralized grid is not unprecedented. Electrical systems were originally dispersed and only later gave way to central production and licensed monopoly. He points out that the decentralized roots from which our present centralized electric grid systems grew are taking on a new strength because of technological progress in small-scale technologies and electronics.

Morris traces the swing of the pendulum toward fewer, larger power facilities and companies, from the heyday of cogeneration through the increasing participation of governments in the regulation of the electric industry to the solid establishment of power pools that make the electric utility effectively a national and even an international enterprise. He points to what he sees as the signs of change toward a decentralized grid.

THE CITY AS REGULATOR AND PRODUCER

The size of power plants has largely determined the organizations involved in the generation and distribution of electricity.

Electric power initially was generated on site. In 1878 Thomas Edison pioneered a new concept. Rather than selling power plants, he would sell electricity. He directed his enormous abilities to this task and was later to reminisce:[1]

> A complete system of distribution for electricity had to be evolved, and as I had to compete with the gas system, this must be commercially efficient and economical, and the network of conductors must be capable of being fed from many different points. A commercially sound network of distribu-

Copyright © 1980 by the Institute for Local Self-Reliance. All rights reserved.

tion had to permit of being replaced under or above ground, and must be accessible at all points, and be capable of being tapped anywhere.

I had to devise a system of metering electricity in the same way gas was metered . . . ways and means had to also be devised for maintaining even voltage everywhere on the system. . . .

Over and above all of these things many other devices had to be invented and perfected such as devices to prevent excessive currents, proper switching gear, lampholders, chandeliers, and all manner of details that were necessary to make a completed system of electric lighting that could compete successfully with the gas system. Such was the work to be done in the early part of 1878.

Edison modestly adds, "The task was enormous, but we put our shoulders to the wheel and in a year and a half we had a system of electric lighting that was a success."

While Edison conceived the idea of retailing electricity, Holly Bridsill developed the first retail heat system. He discovered a way to pump water through pipes under pressure, thus triggering an idea for using steam to distribute heat. He constructed a boiler in the cellar of his house and ran steam through iron pipes to neighborhood houses. When the experiment worked, Bridsill organized the Holly Steam Combination Company. The first successful district heating system was established in 1877 in Lockport, New York.[2]

Edison's plan, while brilliantly conceived, relied on a technology incapable of meeting his requirements. Direct current leaves the power plant at the same voltage that enters the house. The requirement of low voltage at the household level meant that the economic distance electricity could be transmitted in the early days was only a mile or two. Utilities were neighborhood affairs. A few of the contracts between the Edison company and the first major customers, city governments, illustrate the localized nature of the technology. The contract with Washington, D.C., provided for eighty-seven public lamps "burning all night, every night." Wichita paid for seventy-five lights, "located at street intersections." Sacramento obtained thirty-six lights, also for intersections.[3]

Despite Edison's dream, the first major producers of electricity were not the electric utilities but the transportation utilities, the traction (trolley) companies. Within a year of the opening of Edison's Pearl Street power plant in downtown Manhattan, Frank Sprague started the first electric railway in Sarasota Springs, New York. The electric streetcars, or trolleys, as they came to be called, spread across the nation almost immediately. By 1890 51 cities had some electric streetcar service.[4] By 1895 electric trolleys operated in 850 cities over 10,000 miles of track. By 1902 only 665 miles of street railway track (out of a total of 23,577) were not electrified.[5]

At the turn of the century more than half the nation's electricity was generated on site, mostly by the streetcar companies. The technology appeared to allow competition, and cities awarded franchises to many companies. In 1880 the Denver City Council granted a general electric franchise "to all comers."[6] New

York City awarded six franchises in a single day in 1887.[7] Chicago had more than twenty-nine electric companies operating at one time or another in the late nineteenth century.[8]

Cities quickly learned that maintaining competition in this industry did not improve the quality of service, and that it was a difficult proposition in an age characterized by mergers and trusts. The experience of Houston was typical.

> City government had found few effective means to regulate these firms after twenty-five years of experimenting with various franchise strategies. While the introduction of competitive ventures had produced immediate rate reductions and service extensions, businessmen rapidly merged their enterprises to restore monopolies. The consolidation of the two railway companies, as well as the electric firms, by outside investors in 1891 convinced many Houstonians that the maxim "competition is the life of trade" was a misleading myth.[9]

City officials came to recognize that the electric utility was a "natural monopoly," requiring an enormous amount of capital invested in the distribution system before any revenue was generated. Redundant distribution systems would be socially wasteful. Initially the argument for monopoly concerned only the distribution system, but with the rise of the larger steam turbine generators it encompassed generation as well.

Electric utilities would be monopolies. But under whose ownership and control? The poor service provided by the private utilities brought to power a wave of city officials favoring public ownership. "In city after city, referendums favored municipal ownership, and in city after city, advocates of municipal ownership were voted into office."[10] In 1896 there were 400 municipally owned electric plants in the United States.[11] The vast majority were in small cities, which the private utilities found less profitable to service. A decade later there were more than 1,250 publicly owned plants. The rate of increase was more than half again as fast as the rate of increase of privately owned plants. From 1902 to 1907 the rate of increase was more than twice as fast.[12]

But the fiscal crisis of 1907, precipitated by New York City's default on its bonds, caused small cities to lose their bond ratings. Because of their inability to finance their electric systems, the municipal ownership movement lapsed.

At least as important as the depressed financial situation of cities was the evolution of the modern steam turbine. In 1886 the Westinghouse Electric and Manufacturing Company began commercial production of alternating-current machinery on a large scale. This made possible a great increase in voltage capacity and enabled operators to transmit electric energy by means of transformers and high voltage lines over large areas. (It also permitted the water power industry to become the hydroelectric industry.)

In 1901 the Hartford Electric Company became the first American utility to install a steam turbine, which delivered 2 MW.[13] In 1903 Samuel Insull, then president of Chicago Electric (later Commonwealth Edison), installed a 5-MW system, which was dwarfed eighteen months later by a 10-MW system, ten years later by a 35-MW system, and after another ten years, by a 175-MW system.[14]

The steam turbine also began to provide heat as well as electricity. Initially district heating systems used boilers to produce steam that was distributed in pipes laid under the streets and connected to buildings to provide heat and hot water. By the turn of the century several steam turbines produced electricity and heat simultaneously. Between 1900 and 1940 district heating that utilized cogeneration and steam distribution grew rapidly in many northern cities. For the most part these systems supplied the central business districts of such large cities as New York, Philadelphia, Detroit, Boston, and Indianapolis. Development was limited to central areas because steam could not be transported more than three to six miles before condensing as a result of drops in temperature caused by customers using the heat.

The size of steam turbines allowed Samuel Insull to urge those producing their own electricity from smaller plants to abandon them and purchase their power from a central power plant. The large plants could produce electricity less expensively. The greater the number of people hooked into the grid, the greater the "diversity factor." Insull used a block of houses in Chicago's North Side as his favorite illustration.

> There were 193 apartments on that block and 189 of them were customers of the Chicago Edison Company. There were no appliances, motors, or other electrical devices to speak of in that block of dingy apartments—just electric lamps. The power demanded by all the separate apartments on the block, if totalled, was 68.5 kilowatts.
>
> But . . . the different lamps would be lighted at different times, and the actual maximum demand for power from that block of apartments was only 20 kilowatts.
>
> To supply all of these customers from a single source would therefore require generating power of 20 kilowatts. But if each household were to be equipped with a separate generating plant to meet its own needs, an aggregate of 68.5 kilowatts would be needed—more than three times as much.[15]

The argument was sound. Industry and the streetcar companies began to abandon their own power supplies. In 1899 slightly less than 200,000 horsepower in the manufacturing sector was driven by purchased electric energy. Ten years later the figure was ten times that amount. Three out of every five kilowatt hours generated in America came from electric utilities.[16]

Technical improvements in the generation and distribution system permitted electricity to be transmitted over long distances. In 1892 the Southern California Edison Company opened transmission lines for the delivery of 10,000 volts to a point 28 miles away. In 1902 electric power was being transmitted as far as 200 miles in the San Francisco area.[17] The technology had grown beyond the capacity of local government to control it.

THE GROWTH OF STATE AND FEDERAL REGULATION

The battle over which level of government should regulate electric utilities was waged for two decades. To many, such as Delos Wilcox (a public utility expert and author), the issue was intimately associated with the issue of municipal home rule. In 1913 the Committee on Franchises of the National Municipal League concluded that

> the control of all public functions should be localized as much as possible. In this way only can the active and intelligent interest of the voters be aroused and maintained, and the entire machinery of government be kept close to the people for whose benefit it has been created. Also, in this way only can advantage be taken of the local knowledge and the local interest which, given a sound public opinion, are of incalculable benefit in public administration.[18]

But even Wilcox had to concede that "the most serious defect [in local regulation] . . . was found in the fact that the incorporated city or village was no longer the natural unit of control as it ceased proportionately to be the natural economic unit of supply. . . . Public utilities, although still comparatively simple industries, had grown far enough beyond merely local boundaries to require complex governmental machinery to operate or regulate them."[19] In 1907 Wisconsin, New York, and Georgia established state regulatory commissions. By the 1930s most of the states had done the same.

Even today these regulatory commissions vary widely as to their control over municipal electric utilities.[20] Municipally owned utilities are subject to the general jurisdiction of the public utility commissions in nine states (Maine, Maryland, Nebraska, New York, Oregon, Rhode Island, Vermont, West Virginia, and Wisconsin). In others, such as Colorado, Kansas, Mississippi, Pennsylvania, South Carolina, and Wyoming, utilities whose service extends beyond the municipal borders are under state jurisdiction. In Illinois, city governments may regulate the local operations of public utilities if the electorate so chooses at a referendum. In Kansas, local governments have been authorized to regulate public utilities that operate in a single municipality. In New Mexico and South Carolina, local governments are authorized to establish rates to be charged within their borders. All actions are subject to review by the state regulatory commission if there is a complaint.

State regulatory procedures encouraged large power plants and electrical growth in general.

> A fair return on investment seems like a minimal sort of guideline. . . . In practice, the effect was to permit power companies to charge rates sufficient to pay for the physical plant that they built regardless of the motive or prudence of the construction; and to encourage new and still larger investments. . . .
>
> Large power plants could be built, even though they would be only par-

tially used; existing customers would pay for the expansion while the huge new surplus capacity would be used to solicit new customers for large blocks of power at very low prices. Once the power plants were producing as much as could easily be sold, the cycle would be repeated.

The state regulation therefore fueled and almost guaranteed rapid expansion of the power business. . . . The state would allow a fair return on investment, no matter how large; a company, being a monopoly, could charge whatever the state would permit. The more expensive a company's plants, therefore, the more it could charge. Expensive generating plants would expand the profits allowed to a company in absolute terms . . . the company would make more money, but it would not necessarily grow more efficient.[21]

Electric utilities in the 1930s spilled over state borders as easily as they had previously overflowed the boundaries of cities. By 1935, 20 percent of the nation's electrical energy crossed state lines.[22] In 1928, when forty-one of the states either imported or exported more than 19 percent of their power, twenty-two states imported or exported more than 25 percent of their power and seven states imported between 50 and 75 percent of their power.[23]

Utilities began to establish power pools to coordinate electric planning on a multistate basis. The first power pool was established in 1927 by agreement between the Public Service Electric and Gas Company of New Jersey and the Philadelphia Electric Company. The number of power pools increased from four in 1960, representing 12 percent of the nation's capacity, to seventeen in 1970, representing 50 percent of the nation's capacity.[24] These power pools undercut local and state authority even more. The Berkshire County Regional Planning Commission declared, "The size of power pools and the fact that they extend beyond traditional regulatory jurisdictions have created difficulties for representation of local and regional viewpoints."[25] In the 1930s federal agencies became the primary supervisors of these new power systems. The Federal Power Commission (later to become the Federal Energy Regulatory Commission) had authority over any interstate transfer of electricity. At first courts examined, with enormous complexity, what portion of the electricity generated by one utility was in interstate commerce, but with the rise of power pools this became impossible and the courts gave the Federal Power Commission authority if any interstate electrical transfer existed.

PRIVATE AND PUBLIC: STRIFE AND PARTNERSHIP

Many of the cities that owned their electric systems began to relinquish their power plants, much as industries and transportation utilities had done earlier. In 1935 almost half the municipally owned electric utilities generated all of their own power. In 1975 only one in ten did so.[26] In 1978 the United States electric utility industry consisted officially of 3,500 systems, but 2,400 of them were involved solely in the transmission and distribution of power.[27]

After an exhaustive analysis of the electric utility system, the authors of the Berkshire County report concluded, "Centralized power . . . makes local communities dependent on large electric utilities or impels them . . . to purchase a modest share of a large power plant."[28]

The dependency was a precarious one, for private utilities had never accepted the right of municipally owned utilities to compete with them. Private utilities divided up entire areas of the country on an exclusive basis. In one case before the Federal Energy Regulatory Commission (FERC), several Ohio cities argued that the Ohio Edison Company (their wholesale supplier) had territorial agreements with neighboring large private electric utilities. One witness in the case stated, "This means that the cities would have been unlikely to find anyone other than Ohio Edison to sell bulk power to them, even if they could have arranged wheeling over Ohio Edison's lines."[29] The Fifth Circuit Court of Appeals found that the Florida Power Coop and the Florida Power and Light Company were part of a conspiracy to divide the market in Florida.[30]

Private utilities also refused to wheel electricity from lower-cost suppliers to cities. The borough of Grove City, Pennsylvania, contended that the Pennsylvania Power Company refused to sell them wholesale power unless the borough entered into a contract agreeing not to resell the power to industrial and commercial customers.[31] In a proceeding before FERC, twelve Michigan cities charged the Consumers Power Company with refusing to offer interchange service and power supply coordination and transmission services. The city of Breese, Illinois, and six other Illinois cities charged that the Illinois Power Company refused to provide firm wholesale power except on restrictive terms and conditions. They maintained that under these conditions the cities' generating facilities would be virtually useless, or they would be forced as a practical matter to purchase all their electricity from Illinois Power.[32]

The Supreme Court, in the case of Otter Tail Power Company, decided that competition was an important ingredient of our utility system, even if on the local level monopolies were all right. But that decision was a very narrow one. The small towns of Elbow Lake, Minnesota; Hankinson, North Dakota; and Colman and Aurora, South Dakota, set up a municipal distribution system for electricity when the retail franchise of the Otter Tail Power Company expired. Otter Tail refused to sell the new system energy at wholesale prices and refused to permit its wires to be used to deliver electricity from a low-cost federal reclamation project. The Supreme Court overruled the Federal Power Commission's sanction of this action, indicating:

> The bottleneck principle is applicable to Otter Tail. Its control over transmission facilities in much of its service area gives it substantial effective control over potential competition from municipal ownership. By its refusal to sell or wheel power, defendant prevents that competition from surfacing.[33]

The decision to uphold the towns in their suit was 4 to 3. The chief justice dissented, agreeing with the Federal Power Commission that as a retailer of power, Otter Tail asserted a legitimate business interest in keeping its lines free for its own power sales and in refusing to lend a hand in its own demise by wheeling cheaper power from the Bureau of Reclamation to municipal consumers which might otherwise purchase power at retail from Otter Tail itself.[34]

Yet the issue continued to plague city systems. The city of Norwood, Massachusetts, brought suit in the mid-1970s against Boston Edison and the New England Power Company, alleging that Boston Edison refused to provide wheeling services to permit the city to purchase power at wholesale from the New England Power Company.[35] The city of Batavia, Illinois, alleged that Commonwealth Edison Company had prevented the city from acquiring alternate sources of wholesale power by refusing to provide transmission service at reasonable rates.[36] Cleveland's municipal system and a group of Ohio municipal utilities were allocated inexpensive hydroelectric power by the Power Authority of New York as preferred customers. Pennsylvania Power and Light agreed to wheel the power from New York State, but Cleveland Illuminating refused to transmit the power over its lines from the point of interconnection with Pennsylvania Electric.[37]

Municipal systems began to develop their own wholesale agencies. By 1979 the Missouri Basin Municipal Power Agency consisted of fifty municipal electric utilities, and its members crossed the borders of Iowa, Minnesota, North Dakota, and South Dakota. The Arkansas River Power Authority in Colorado encompassed five Colorado cities and one New Mexican city. As the technology increased in size, municipals and private utilities began to form partnerships, called joint action agencies, to finance these systems. Municipal systems could no longer afford to purchase power plants large enough to provide them with cheap electricity, but they could buy a piece of a plant. Private utilities could then gain access to the tax-exempt bonds that utilities owned by municipalities could issue. The interest rate on these bonds is 2 to 4 percent lower, on average, than that on investor-owned utility bonds.

By 1979 the Municipal Electric Authority of Georgia owned a 17.7 percent share in each of two operating Georgia Power Company nuclear plants, and a 15 percent share in each of two Georgia Power Company coal-fired plants.[38] North Carolina's Municipal Agency #1 held a 75 percent interest in a 1,100-MW nuclear plant operated by Duke Power Company.[39] The Massachusetts Municipal Wholesale Electric Company, representing forty municipal utilities in that state, owned a 12 percent interest in the Seabrook nuclear plant, in partnership with the Public Service Company of New Hampshire.[40]

The largest of the new power plants were nuclear reactors. The environmental and security considerations associated with the 1,200 large-scale nuclear power plants that were projected to come on line by the year 2000 led the federal government to preempt local and state authority over siting. In *Northern States Power Company v. Minnesota* the Supreme Court ruled that Congress had preempted by "implication" the authority of the state to set radiation standards more stringent than those set by the federal government.[41] In *Village of Bu-*

chanan v. Nuclear Regulatory Commission a lower New York court ruled that local zoning ordinances were effectively preempted by federal regulations relating to nuclear power plants (a decision that was subsequently modified on appeal).[42]

By the dawn of the energy crisis America was heavily committed to electrification as a way of life. In 1930 one in every ten units of energy consumed in the country was used to generate electricity.[43] By 1960 one in five was used in this manner.[44] In 1980 the proportion was almost one in three.[45] Several studies predict that by the end of the century more than half of the primary energy used in the United States will go for electricity.[46] In fact, because most of the primary energy used in this process is expended as waste heat, the most rapidly growing component of the United States energy budget in the 1970s was waste heat.

Cities, their businesses, and neighborhoods had become no more than bit players in the energy drama. Large industrial corporations that had generated their own electricity were fearful that the federal government would regulate them as utilities if they sold this electricity to the grid systems. Utilities often either refused to interconnect with small power producers or charged steep backup prices for this right. By 1975 less than 5 percent of the nation's electricity was generated on site.[47] The few municipal electric systems that had any generating capability were joining in consortia with private utilities to buy a piece of a huge coal or nuclear project. The boundary between private and public in these consortia had become so blurred that the Internal Revenue Service was forced to issue a regulation in the mid-1970s that tax-exempt bonds could not be used if more than 25 percent of the electricity was going to be used by private utilities.[48] As the nation enthusiastically embraced nuclear power, the environmental impacts of this source of power and the national security implications caused states to preempt local authorities, and the federal government to preempt states.

Then came the oil embargo of 1973 and the 400 percent price hike in crude oil by OPEC. By 1980 the price of crude oil was fifteen times higher than in 1973. Deregulation of domestic petroleum and natural gas supplies in the United States meant that by the mid-1980s these would reach world price levels as well.

These unprecedented price increases undermined the assumptions underlying conventional electric generating systems in America. They made traditional technologies obsolete overnight, while encouraging the development of other technologies more compatible with the era of dwindling cheap fuels. These technologies, in turn, changed the conventional regulatory and ownership structures evolved during the previous century. Cities, industries, and even neighborhoods and apartment houses were once again to play a major role in our electric system.

THE NEW AGE: SMALL IS BEAUTIFUL

As electricity prices rose, utility economists discovered to their chagrin that electric demand was "elastic." As prices rose, demand fell. People began using electricity less, or more efficiently. The 6 to 7 percent annual increases of the 1950s and 1960s gave way to 1 and 2 percent growth rates. For example, the Edison Electric Institute had predicted a 6 percent growth rate for 1979, while the actual peak growth rate was less than 1 percent.

The very scale of power plants exacerbated the impact of this reduction in demand. A 7 percent annual increase in demand leads to a doubling every decade. A 2 percent annual increase will double demand every thirty-five years. It requires a long lead time to build very large power plants. Utilities guessed wrong in the late 1960s and early 1970s and were saddled with a huge excess capacity in the late 1970s. Idle or underused power plants cost millions, even hundreds of millions of dollars in carrying charges.

To minimize their financial risk, utilities began to emphasize smaller power plants which could come on line more rapidly, matching changes in demand more closely. They found that it was much more difficult to raise capital in the money markets for plants that would not return on the investment for a decade. Small power-plant investments could repay more rapidly.

Utilities discovered that several small power plants could replace one larger plant. The overall system reliability increased. Since the backup power required to meet potential forced outages is based on the size of the largest generator in the system, smaller power plants required smaller reserves.[49]

The efficiency with which electricity was generated took on a new urgency. Power plants built in 1975 were no more efficient than those built in 1950. More than 70 percent of the primary energy was lost in generation, transmission, and distribution. Yet by capturing the waste heat, overall efficiencies of "cogenerators" can range from 75 to 85 percent. As the nation looked to cogeneration as a source of power, it became economical for power plants to locate close to their customers. As power plants increased in scale after World War II, they moved further from population centers, reducing the district heating potential. In 1978 the International District Heating Association noted that there were forty-four steam district heating utilities operating in the United States and that there had been a "general decline in the industry with a decrease in steam sales of about 6 percent from 1976 to 1978."[50] U.S. district heating systems using steam distribution in the central core of large cities and some hot water distribution on college campuses supplied the heating requirements of a total of about 2.5 million people, or about 1 percent of the American population. Suddenly, industries that had generated their own power a century before looked again at this possibility.

The steep price rise in crude oil, coupled with federal tax incentives, brought renewable energy technologies into the marketplace. Initially these technologies were competitive only with the production of heat by electricity. But by 1980 solar electric systems, such as wind turbines and biomass-fired power plants, were competitive with conventional generators in some cases.

The scale at which cogeneration and solar electric systems became competitive dropped as energy prices rose. In the 1950s only very large hydroelectric plants, serving hundreds of thousands of customers, were competitive with petroleum-fired plants. By the mid-1970s low-head, or small-scale, hydroelectricity was coming into its own, opening up thousands of sites abandoned earlier, when cheap petroleum was discovered. By 1980 the term *micro hydro* had become part of our vocabulary. It refers to a system that might serve only a dozen houses and use no dams, but there are tens of thousands of potential sites.

Several manufacturers looked at the mass-production capabilities of the automobile engine plants and redesigned the engines to produce electricity and heat. Small cogeneration systems serving fewer than a dozen houses came into the marketplace in Europe.

In 1978 the federal government formalized the new realities of electric power generation in the Public Utilities Regulatory Policies Act (PURPA). Section 210 opened the grid system to all qualifying producers of electricity by cogeneration or renewable sources. Utilities were required to purchase this electricity at the same price they would have had to pay to bring an additional unit of electricity on line. The congressional committee made clear that the cost of "avoided" power must be viewed over a long term.

> In interpreting the term "incremental cost of alternative energy" the conferees expect that the Commission and the States may look beyond the cost of alternative sources which are instantaneously available to the utility. . . . For example, an electric utility which owns a source of hydroelectric power and which is offered the sale of electric energy from a co-generator or small power producer might, if measured over the short term, have a low incremental cost of alternative power because of its access to hydropower; however, it may be the case that by purchasing from the co-generator or small power producer and saving hydropower for later use, the utility can avoid the use of expensive electric energy generated by fossil fired units during later months of its seasonal generation cycle. Thus, viewed over the longer period of time, the incremental cost of alternative electric energy might be substantially higher than that measured by the instantaneously available hydropower.[51]

Utilities were prohibited from putting obstacles in the way of small power producers. They were encouraged to wheel electricity across their grid system so that a qualifying facility could sell electricity to another utility if it could get a better price. PURPA permitted qualifying facilities to bargain collectively with the utility for a better price. Finally, PURPA exempted qualifying producers of electricity from state and federal regulatory, licensing, and siting procedures, applying to both public and private utilities.

The new and resurrected technologies, the new economics of electric generation, and PURPA are effecting a profound revolution in our electric system. For several reasons cities have been the first to feel the impact. Departments of municipal government that had never considered themselves in the energy business are becoming producers of electricity. In Los Angeles and Portland, Oregon, water and irrigation authorities are exploring the potential for hydroelectricity. In several cities in Ohio methane gas from sewage systems is used to generate electricity that is then sold to the private utility. Many cities look at their growing mounds of solid waste, and diminished landfill space, and begin to burn the wastes. Hempstead, Long Island, generates electricity and sells it to the private utility. (However, a small but influential group of critics protests this practice, pointing out that the energy embodied in solid waste, that is, the energy used to process the raw material into the final product, is worth more than the direct Btu content. Recycling, they assert, not combustion, should be the watchword. Burning plants should be sized only for the portion of the waste stream that is not recyclable—about 20 percent.)[52]

In this new era the boundary between gas and electric utilities is beginning to break down as cogeneration grows in popularity. For example, the California Public Utilities Commission in 1979 required its largest utility, Pacific Gas and Electric (PG & E), to sell gas to cogenerators at the same price it charges its own electric division. Yet the economics of cogeneration will depend on the price received from the PG & E electric division. This might produce organizational schizophrenia. One analyst in the financial department of PG & E agreed that some traditional sections of the utility might be opposed to the "subsidization of competing units, but people in the financial planning department believe that it is now inimical to our interests to have to build additional capacity at higher and higher costs."

In New York City no such schizophrenia need exist. Consolidated Edison, which holds the natural gas franchise for Manhattan, refuses to connect up cogenerators, while Brooklyn Union Gas experiments with household cogeneration systems in its service area.

The New England and Middle Atlantic states have been rediscovering the town river as a source of power. Franklin Falls, New Hampshire, has ten existing dams. Auburn, New York, has six. Who will own these facilities? In Franklin Falls the first two are owned by local entrepreneurs. In Springfield, Vermont, municipal ownership appears to be the solution. In Lawrence, Massachusetts, a limited partnership was established for rich investors to gain access to the lucrative tax benefits for hydroelectric power. In New York State a private entrepreneur and a huge insurance company are prospecting for hydroelectric sites.

In many cases even if preference is given to public ownership, it is unclear what the level of government will be. In New Hampshire the state owns the dams, but it gives or leases them to anyone who wants to develop them. But the state water resources board and industrial revenue authority have found it difficult to know whom to give jurisdiction to. New Hampshire law makes it difficult for towns to own and operate their own electric generation facilities.

The Federal Energy Regulatory Commission has the authority to confer exclusive licenses for a hydroelectric site. The borough of Lehighton, Pennsylvania, in 1980 filed an application for a permit to study a proposed hydroelectric project at the Army Corps of Engineers' Beltzville Dam on the Pohopco Creek in that state. The Delaware River Basin Authority and the state of Pennsylvania filed competing applications for the same project.

But the greatest controversy has involved the growing number of inner-city organizations that have begun to generate electricity through cogeneration, using petroleum or natural gas as their fuel source. Cities, with their high population densities, offer the most attractive site for cogeneration. On the other hand, local utilities, especially in the East, have fought hard against these new intruders, despite the PURPA regulations.

For example, Harvard University had gotten into the power business when its medical school moved to its present location on Longwood Avenue.[53] To supply its new buildings with steam and electricity the university built a powerhouse. Later, when air conditioning swelled demand for chilled water, the plant relinquished the generation of alternating-current electricity to the Boston Edison Company but continued to supply its customers with compressed air, oxygen, and some direct-current electricity. By the 1960s the energy demands of the medical area threatened to outstrip the capacity of the aging powerhouse. Boston Edison was not interested in forming a partnership with Harvard. In 1972 the Medical Area Service Corporation was established to plan and coordinate services. Harvard installed what was to be the nation's largest venture in cogeneration, a 73-MW system that would simultaneously supply steam, chilled water, and electricity to a ten-block enclave in the Fenway–Mission Hill district of Boston.

However, the surrounding neighborhood, Boston Edison, and later the nearby town of Brookline opposed the plant because of increased NO_2 emissions. The Massachusetts Department of Environmental Quality eventually permitted the system to generate steam, but not electricity. After five years the cost of the plant had surpassed its estimate by several hundred percent and was the largest financial undertaking in Harvard's history.

In New York City the Seal-Kap Company, a packager of lids for Dannon Yogurt and other companies, tried to remain in the area despite very high electric rates.[54] In order to save its 180 local jobs, and more than $200,000 a year paid in New York taxes, it installed diesel-fired generators. Con Edison opposed the installation "for the same reason we oppose all on-site generators," said the successor to Thomas Edison's first Pearl Street power station. It charged Seal-Kap steep standby rates, equal to 75 percent of Seal-Kap's recent utility bills, whether it consumed this much electricity or not. Seal-Kap reacted by installing an extra diesel generator as a back-up system and uncoupled completely from Con Edison. This action was four times less expensive than continuing to interconnect with the high back-up charge. Within a month Seal-Kap was paying a nickel rather than a dime for one kWh of electricity. However, Con Edison persuaded the city of New York to change the tax basis of Seal-Kap's cogeneration plants to real property. By July 1980 Seal-Kap found that its city taxes were $3,000 a month higher than it had anticipated. Coupled with rising petroleum prices, the pay-back period for the plant had doubled to six years.

Perhaps one of the most intriguing examples of the new era occurred in some New York City cooperative apartment complexes. Five apartment complexes in 1980 formed a non-profit corporation, called the National Urban Energy Foundation Corporation, to help users obtain loans to purchase and install cogeneration equipment. Utility costs at that time accounted for more than 50 percent of most apartment operating budgets, up from 10 percent in 1970.[55] According to Murray Raphael, president of one member complex, Luna Park, cogeneration had become a "do or die" issue. "Our utility [Consolidated Edison] has become a major destabilizing factor in the budgets of thousands of financially strapped middle income housing complexes in the city," said Richard Stone, founder of the corporation and manager of the 1,000-unit Big Six development in Woodside, Queens. Stone had a larger ambition for the corporation: he wanted to create a massive energy cooperative, not only to finance conversions but also to organize electrical distribution among its members and provide its own fuel supplies. In August 1980 Stone placed a bid with New York City's Department of Sanitation for drilling rights to a landfill dump in the Pelham Bay section of the Bronx. A city prospectus estimated the dump's productivity at four million cubic feet of methane per day. In June 1980 Big Six had switched to cogeneration and, like the Seal-Kap corporation, reduced kilowatt hour costs by 50 percent, from ten cents to five. "We hope to get down to three or four," said Stone. "Those kinds of figures are attractive to a lot of potential members."

Not all utilities discouraged on-site cogeneration. In 1976 the publicly owned Eugene Water and Electric Board (EWEB) and the giant Weyerhauser Paper Company entered into a cooperative agreement. The latter planned to install 55 MW(e) of generating capacity at its Springfield mill. EWEB agreed to lease from Weyerhauser the land immediately adjacent to the mill, on which it designed, constructed, owns, and now operates generating facilities dependent upon the steam boiler facilities owned and operated by Weyerhauser as part of its pulp and paper operations.[56] Pacific Gas and Electric, after having a rate increase reduced by the California Public Utilities Commission for its lack of progress in bringing cogeneration facilities on line, is working with San Francisco's General Hospital and, in Oakland, is studying the feasibility of cogeneration for the convention center complex, where heating and cooling would be provided to the office, hotel, and convention center.[57]

City government was only marginally involved in these cogeneration disputes, except around the issue of air pollution. But when advocates of district heating began to convince city officials and the federal departments of Energy and Housing and Urban Development of the feasibility of district heating systems retrofitted to existing power plants, the role of the cities became primary. Few state regulatory commissions exercised authority over the distribution of heat, and several studies concluded that district heating would be technically feasible and economically competitive in many cities.[58] One report concluded:

> The technology is well established. The cost/benefit yield is favorable, and the conservation potential is significant. Application of district energy should occur in urban and densely populated suburban areas. The remaining portion of the space conditioning and domestic water heat demand located in rural and low population-density communities appear to be better suited to other forms of system distribution.[59]

In order to better assess the role of local governments in district heating, many experts visited Europe and Japan. One observer concluded:

> The degree of implementation of district heating in Europe appears to be in direct proportion to the degree of involvement of the government in providing policies regarding energy use. The same rules apply in the U.S. . . . District heating would surely compete with conventional service industries, and eradicate many heating oil distributorships and considerably reduce natural gas sales in urban areas. District heating may become a new utility, separate from those now in existence, especially if developed municipally, or it may represent a growth area for existing utilities. In any case, ownership, licensing, and regulation must be clarified. This includes definitions of system reliability criteria and the liability of industries to supply heat upon demand, and establishment of criteria for cost allocation particularly between electric rates and heat rates. This may be further complicated by the growing practice of power pooling and economic dispatch of power, since heat demand may necessitate operation of an inefficient plant.[60]

Initially the new producers of electricity were allowed to sell their electricity only to utilities. But New Hampshire amended its Limited Electrical Energy Production Act in 1979 to permit qualifying facilities to sell electricity at retail to up to three customers. Harvard hoped that once it could begin generating electricity it could wheel electricity across the grid system to other university facilities. So did the apartment complexes in Manhattan.

PHOTOVOLTAICS COME OF AGE

To many people, the confused roles of the late 1970s and early 1980s would pale to insignificance in the light of the massive changes that would come once photovoltaics became inexpensive enough for widespread application. Solar cells, as they are popularly called, have been used to power satellites since the late 1950s. Since 1973 they have been used on earth, primarily for remote applications where long-distance wires had to be brought into the area, or heavy-duty batteries were the alternative. Irrigation systems, highway warning lights, water-buoy lights, oil rigging platform lighting, radio repeater stations, even an outhouse at the Yellowstone National Park were operating with solar cells by 1980. By 1985 the price could decrease more than tenfold, making this technology competitive for household applications.

As with conventional power plants, solar cells are more attractive if they use solar cogeneration. Most solar cells are made from silicon. At high temperatures the cell efficiency drops. By taking off the excess heat, higher operating efficiencies are achieved, and more than twice as much overall useful energy per square foot of collector area. One company, Selectrothermo of Dracut, Massachusetts, was offering thermal/electric systems in 1980, and with federal tax incentives it had a seven-year payback.

About five hundred to one thousand square feet of photovoltaics would provide all the electrical and thermal requirements for an average single-family detached household by 1985. However, many people, knowing that two-thirds of the population of this country resides in urban areas, err in thinking that there is insufficient space in our cities for on-site generation from direct sunlight.

As one historian notes:

> The twentieth century has been a period of strong population deconcentration within American metropolitan areas. Population deconcentration has been consistent across metropolitan areas and systematic over time. . . . At the turn of the century a city of 100,000 was likely to be concentrated in an area of ten to twenty square miles. Population density at the core of the city would have been relatively high. . . . As a contrast, contemporary Los Angeles and San Diego have rather even population densities of three to five thousand people per square mile spread over hundreds of square miles.[61]

The highest densities occur in large cities. Yet of all cities with over 100,000 people, the average density is only 4,480 people per square mile.[62] Densities range from New York City's 26,343 residents per square mile to Oklahoma City's 577. Single-family homes comprise two-thirds of all housing structures in urban areas.[63]

Decentralized applications of photovoltaics, however, depend not only on total city density but on the density in specific communities within municipal borders. The city as a whole may have a low density, but its commercial and residential areas may be densely populated. Two 1950 studies examining Dayton and Cincinnati just before the suburban exodus and urban renewal programs began to thin out populations in the central cities should represent worst cases in that they are older, industrially based cities. Within one mile of the city center both Dayton and Cincinnati had population densities of 20,000 to 80,000 people per square mile, or about 20 to 120 people per acre. Between one mile and three miles of Dayton's center, densities drop sharply to 5,000 to 30,000 people per square mile. In Cincinnati the drop was even more abrupt. Outside of the three-mile ring the densities begin to equal those of the overall large-city average, 4,460 per square mile, or about 8 people per acre.[64]

How much of the south side of a structure can be exposed to solar energy in our compact cities? Using a concept called a solar envelope, which is the largest volume in which a building will not shade adjacent parcels, Professors Knowles and Berry at the University of Southern California demonstrated that quality, moderate-density development is achievable while protecting solar access. Averaging results from six different sites, a density of 52 dwelling units per acre was achievable.[65] Assuming an average of 2 people per unit, the density of 66,000 people per square mile is considerably higher than that of our largest American cities.

The Jet Propulsion Laboratory evaluated the potential for rooftops in the San Fernando Valley in California to supply household energy from photovoltaics. Nevin Bryant, the author of the report, concluded, "For the sixty-five square mile study area, the results showed that with half the available flat and south facing roofs used and assuming the availability of energy storage, 52.7 percent of actual energy demand could have been met in 1978 using photovoltaic collectors."[66]

The Urban Innovations Group at the University of California at Los Angeles compared three solar urban futures for a city of 100,000.[67] It concluded that the residential sector could be totally self-sufficient if 80.7 percent of available residential roof area is used. The commercial sector could collect 67 percent of its energy demand by using about 50 percent of available parking area and 100 percent of available rooftops. The industrial sector could collect 18 percent of its energy needs on site. The study concludes, "However, if land area in the hypothetical city is increased 34.5 percent (from 10,000 to 13,450 acres, or from a gross density of 10 persons per acre to approximately 7.4 persons per acre) all three sectors could be energy self-sufficient."

Finally, the Department of Energy concludes:

> For grid connected PV [photovoltaic] applications, sufficient roof area will exist in the residential sector to satisfy residential electrical energy requirements when the sun is shining. And, on the average, sufficient roof area will exist to supply a large percentage of the total electrical energy requirements. In fact, if 10 kWp capacity per dwelling is installed on a large fraction of existing homes, the residential sector can become a net exporter of electricity.[68]

DOE is less sanguine about the possibilities of the commercial and industrial sectors' becoming self-sufficient.

> Although exceptions will exist, in general the intermediate load centers (commercial and industrial sectors) will not be able to meet their electrical demands by on-site PV because of lack of available adjacent land. This will be particularly true in most urban areas. Not only will PV systems not be able to provide a large percentage of intermediate load requirements on-site, they will also be constrained in the amount of instantaneous load they can supply. In these situations the grid could be used to supply a greater percentage of the total electrical load than in the residential or remote sectors.[69]

An intriguing and complex dynamic may occur once household electric systems become a reality. We can expect households to maximize their rooftop areas, especially if they will be able to generate revenue from sales of electricity to the grid. The household electric vehicle will probably become part of the residential load. Assuming a car consumes .5 kWh for each mile driven, the transportation electrical load would be approximately equal to that of a refrigerator and lighting today.

Initially, when the federal government analyzed the economics of dispersed photovoltaics it assumed that they would be owned by utilities, which would pay a roof fee to the homeowner. However, as an MIT energy laboratory report indicated, "their analysis required *utility ownership* of the systems, and thus the framework in which their financial analysis was performed failed to capture many of the potential advantages of residential, user-owned systems."[70] The possible conflict of interest between the on-site generator and the utility company should be considered. DOE notes that photovoltaic power systems "add considerably to the reliability of the consumer's supply of power, though they may not be adding to the overall reliability of the system." For illustration it uses a nonurban example:

> A case in point is the experience gained in the Meade, Nebraska, agricultural p/v experiment . . . during the first 15 months of operation of the Meade system, power from the utility has been unavailable 47 times. For the most part, these interruptions have been caused by failures in the distribution system, in which the Meade station lies at the end. During the same period of time, the photovoltaic power system was unavailable for power production only 15 times. Thus, with sufficient electrical storage, a p/v system can be considerably more reliable at the customer end than with the utility alone.[71]

The user may value reliability differently than the utility. Utilities currently try to maintain enough capacity to ensure that failure of the generating plant will curtail power no more than 2.4 hours per year. Southern California Edison uses a standard of 1 hour in ten years. It has been argued that this standard for generating reliability is too high, and that the last few hundredths of a percent of reliability are enormously expensive, particularly since the transmission and distribution system is usually less reliable than the generating plant.[72] The Office of Technology Assessment agrees:

> Standards for reliability cannot be measured in any systematic way. Requirements will differ from customer to customer. Some industries, for example, would face catastrophic losses if they lost power for an extended period (say several hours), while residential customers might not be willing to pay a premium for extremely high reliability. One of the disadvantages of providing power from a centralized utility grid is that all customers must pay for high system reliability whether they need it or not. On-site generation would permit much greater flexibility in this regard.[73]

DOE concurs:

> The availability of customer generated p/v power combined with energy storage will offer the option to such a consumer of choosing his own reliability level rather than accepting the level of overall system reliability.[74]

Unfortunately, as DOE admits, "Analysis of the requirements of different types of customers in this regard is almost nonexistent. It is difficult to anticipate how much different customers would be willing to pay for reliability if they were given a choice."[75] What is it worth to people if five days per year they have to postpone washing their clothes, or turn the thermostat lower during the night than they might like, or take shorter showers than usual?

There are, however, attractions to moving beyond self-sufficient households.

The diversity factor that Samuel Insull stressed is still important. The greater the number of consumers linked into the grid, the lower will need to be the per capita generating capacity to serve their diversified loads (those that are not weather related). Although battery storage is possible at the household level, the dangers involved in having many heavy-duty batteries in each household can be reduced by placing them in a central place. The cost of battery shelters is not directly proportional to their size. Also, some of the subsystem components can be more efficient and less costly if larger. The Electric Power Research Institute discusses the disadvantage of small inverters—devices that change the direct current coming from photovoltaics to alternating current usable in contemporary appliances—and concludes that household inverters will have an efficiency of 90 percent compared with a 97.8 percent efficiency for inverters sized in the tens of megawatts range.[76]

Finally, there is the issue of the coincidence of the generation of power by rooftop or ground-mounted solar arrays and the demand. General Electric modeled various array sizes and found that, without storage, there is a significant mismatch between energy generation and demand within the residential household.[77] For example, a house in Boston with solar arrays generates 7.3 MWh per year, but only 4.3 MWh goes directly to the household load. The rest goes to the utility, and the household buys 10.79 MWh from the utility. A Phoenix household shows this effect even more dramatically. An array of 800 square feet generates 15.2 MWh, more than the annual household energy demand of 14.7 MWh. However, because of the insolation/demand mismatch, the Phoenix house exports 8.3 MWh to the utility, slightly more than it imports from it. In Seattle the mismatch is similar. A household with a 1,000-square-foot array would sell almost 7 MWh to the utility and purchase 8.5 MWh from it.

Neighborhood electric cooperatives; district heating systems operating from a municipally owned power plant; rooftop systems integrated into a community storage and backup system; utilities as mere dispatching agents, or maintenance crews, or financing mechanisms; cities that locate new businesses based on the need for waste heat—these are a few of the visions of the urban future touched on here.

No one can know the future. Overnight we have witnessed a dramatic change in the assumptions underlying our electrical generation technologies and regulatory procedures. The change in the price of crude oil has come much faster than the changes in our laws, customs, or institutions. As the society strives to catch up with the new calculus of electrical energy production, we can expect structural tensions. The birth process of the new electric system may be as painful as that of the old. Just as those who built the first power plants in the 1870s could not foresee the organizational structures and regulatory problems that would evolve over the next half century, neither can we forecast with any confidence the strains and stresses that will come with the construction of a new system. We know that it will be much more varied. There will be hundreds, perhaps hundreds of thousands, more producers. Regulation will become more localized. Electric and thermal generation will be accomplished simultaneously, forcing us to regulate the distribution of heat as well as electricity. And people will be much more directly involved in designing systems that match their needs. It is an exciting time, the birth of a new age.

NOTES

1. Edward Hungerford, *The Story of Public Utilities* (New York: G. P. Putnam's Sons, 1928), pp. 148–50.
2. Todd Anuskiewicz and Roy Meador, "District Heating Backgrounder" (Paper delivered at the North Central State Energy Seminars on District Heating, Detroit, May 29–30, 1980), p. 1.
3. Ellis L. Armstrong, ed., *History of Public Works in the United States: 1776–1976* (Chicago: American Public Works Association, 1976), p. 3.
4. Charles N. Glaab and A. Theodore Brown, *A History of Urban America* (New York: Macmillan Co., 1967), p. 92.
5. Ibid.
6. Martin Glaeser, *Public Utilities in American Capitalism* (New York: Macmillan Co., 1957), p. 60.
7. Charles F. Phillips, Jr., *The Economics of Regulation* (Homewood, Ill., 1969), p. 83.
8. Marc Messing, H. Paul Friesema, David Morell, *Centralized Power: The Policies of Scale in Electricity Generation* (Cambridge, Mass.: Oelgeschlager, Gunn and Hain, 1979), p. 20.
9. Michael H. Ebner and Eugene M. Tobin, eds., *The Age of Urban Reform: New Perspectives on the Progressive Era* (Port Washington, N.Y.: Kennikat Press, 1977), p. 33.
10. Forrest MacDonald, "Samuel Insull and Utility Regulation," *Business History Review* 32 (1958):241–54.
11. Ibid.
12. Ibid.
13. EPRI, *Creating the Electric Age* (Palo Alto: Electric Power Research Institute, March 1979), p. 61.
14. Ibid., p. 62.
15. Sheldon Novick, "The Electric Power Industry," *Environment*, November 1975, p. 32.
16. Glaeser, *Public Utilities*, p. 98.
17. Ralph Woods, *America Reborn: A Plan for Decentralization of Industry* (London: Longmans Green and Co., 1939), p. 225.
18. David Nord, "The Experts versus the Experts: Conflicting Philosophies of Municipal Utility Regulation in the Progressive Era," *Wisconsin Magazine of History* 58 (1975):219–36.
19. Glaeser, *Public Utilities*, pp. 120–21.
20. Duane A. Feurer et al., "Study of the Impacts of Regulations Affecting the Acceptance of Integrated Community Energy Systems" (Prepared for the Department of Energy, Contract no. EM-78-C-02-4627, September 1979).
21. Novick, *Environment*, November 1975, p. 32.
22. Glaeser, *Public Utilities*, pp. 120–21.
23. Messing et al., *Centralized Power*, p. 170.
24. Ibid., p. 49.
25. Berkshire County Regional Planning Commission, *Evaluation of Power Facilities* (Springfield, Va.: National Technical Information Service, April 1974).
26. Messing et al., *Centralized Power*, p. 24.

27. Ibid., p. 20.
28. Ibid.
29. Ohio Edison Company Docket no. ER 77-530, Federal Energy Regulatory Commission, Testimony of Ralph E. Muller, April 1978, p. 162.
30. *Gainesville Utilities Department v. Florida Power and Light Co.*, Fifth Circuit Court of Appeals, no. 76-1542, May 22, 1978.
31. *Borough of Elwood City and Borough of Grove City, Pennsylvania v. Pennsylvania Power Company*, Civil Action no. 77-1145, U.S. District Court for the Western District of Pennsylvania, October 3, 1977.
32. *City of Breese, Illinois, et al. v. Illinois Power Company*, U.S. District Court for the Southern District of Illinois, Southern Division, Civil Action no. 78-C-3111, January 30, 1978.
33. *United States v. Otter Tail Power Co.*, 331F. Supp. 54 (1971), Aff'd. 410 U.S. 366.
34. *Otter Tail Power Company v. United States*, 93 S. Ct. 1022 (at 1035), 1973.
35. *Town of Norwood, Mass. v. Boston Edison Company et al.*, Civil Action no. 74-4104-T, U.S. District Court for the District of Massachusetts.
36. *City of Batavia, Ill., et al. v. Commonwealth Edison*, Civil Action Court for the Northern District of Illinois Eastern District, November 29, 1976.
37. The Atomic Safety and Licensing Board, Atomic Energy Commission, Initial Decision, January 6, 1977, pp. 76–78.
38. *Public Power* Magazine, November/December 1979.
39. Ibid.
40. Ibid.
41. Messing et al., *Centralized Power*, p. 74.
42. Ibid.
43. House of Representatives, Select Committee on Small Business, *Promotional Practices by Public Utilities and Their Impact upon Small Business: Hearings before the Subcommittee on Activities of Regulatory Agencies*, 90th Cong., 1968, p. 408.
44. Ibid.
45. Energy Information Administration, *Annual Report to Congress*, 1978.
46. See, e.g., *Alternative Energy Demands Futures to 2010*, Report of the Demand and Conservation Panel to the Committee on Nuclear and Alternative Energy Systems, National Research Council, National Academy of Sciences, Washington, D.C., 1979.
47. Federal Power Commission, News Release, May 6, 1975, p. 4.
48. See IRS regulation 1.103.7.
49. See, e.g., the study of a Los Alamos Scientific Laboratory comparison of the use of small coal-fired power plants with the use of large power plants for expanding electrical generation capacity in the Rocky Mountain West. Reported in *Environment*, March 1980.
50. Cited in Jack Gleason, *District Heating in American Cities*, forthcoming.
51. Conference Report on H.R. 4018, Public Utility Regulatory Policies Act of 1978, H.R. 1750, 98–99, 95th Cong., 2d sess. (1978).
52. See, e.g., Neil Seldman, *Garbage in America* (Washington, D.C.: Institute for Local Self-Reliance, 1979).
53. The following description comes from an article in *Harvard Magazine*, July/August 1980.
54. The following description comes from an article in *The New York Times*, August 24, 1980.
55. The following description comes from an article in *Energy User News*, August 24, 1980.
56. Messing et al., *Centralized Power*, pp. 139–40.
57. Frederick W. Mielke, Jr., Chairman of the Board, Pacific Gas and Electric Company, in a speech published in *Energy Efficiency and the Utilities: New Directions* (San Francisco: California Public Utilities Commission, 1980), p. 88.
58. See, e.g., J. Karkheck and J. Powell, "Waste Heat as an Alternative Energy

Source," Brookhaven National Laboratory Report BNL-23751, December 1977, and M. Loszewski, "Preliminary Investigation of the Thermal Grid Concept," ORNL/TM-5786, Oak Ridge National Laboratory, Oak Ridge, Tenn., October 1977.

59. John Karkheck and James Powell, *The Application of District Heating Systems to U.S. Urban Areas*, BNL-25022, 1980, p. 14.

60. Ibid., p. 13.

61. Barry Edmonston, *Population Distribution in American Cities* (Lexington, Mass.: Lexington Books, D.C. Heath and Co., 1975), pp. 135–36.

62. *Statistical Abstract of the United States*, 100th ed., pp. 24–26.

63. Ibid.

64. Edmonston, *Population Distribution*, p. 57.

65. R. Knowles and R. Berry, "Solar Envelope Concepts: Moderate Density Applications" (Los Angeles: University of California, 1980).

66. Jet Propulsion Laboratory, *Some Currently Available Photovoltaic System Computer Simulation Approaches*, Technical Report 5250–2 (Pasadena: U.S. Department of Energy, July 31, 1979).

67. R. L. Ritschard, *Assessment of Solar Energy within a Community: Summary of Three Community Level Studies* (Washington, D.C.: U.S. Department of Energy, October 1979).

68. *Federal Policies to Promote the Widespread Utilization of Photovoltaic Systems*, vol. 2 (DOE/CS–0114/2, February 1980), p. 11–5–6.

69. Ibid., p. 11–6.

70. Paul R. Carpenter and Gerald A. Taylor, *An Economic Analysis of Grid-Connected Residential Solar Photovoltaic Power Systems* (Cambridge, Mass.: MIT Energy Laboratory Report, May 1978, revised December 1978), p. 21.

71. *Federal Policies*, p. 2–13.

72. Michael L. Tolson, "The Economics of Alternative Levels of Reliability for Electric Power Generation Systems," *The Bell Journal of Economics*, vol. 6, no. 2, Summer 1975, p. 697.

73. *Application of Solar Technology to Today's Energy Needs* (Washington, D.C.: Office of Technology Assessment, September 1978), vol. 1, p. 132.

74. *Federal Policies*, p. 2–13.

75. Ibid., p. 2–14.

76. EPRI, *Requirements Assessment of Photovoltaic Power Plants in Electric Utility Systems* (Palo Alto: Electric Power Research Institute, EPRI ER–685, Project 651–1, June 1978), Full Report, pp. 6–11.

77. EPRI, *Requirements Assessment of Photovoltaic Power Plants in Electric Utility Systems* (Palo Alto: Electric Power Research Institute, prepared by General Electric Company, Schenectady, N.Y., June 1978), pp. 10–110.

PART II

Managing Decentralization: The Tools, Rules, and Fuels of Reform

LISA FRANTZIS

CHAPTER 4 | *The Potential for Diversity: The Production Alternative*

Lisa Frantzis's 1979 thesis at Wesleyan University's College of Science in Society assessed numerous major national studies attempting to quantify the gross potential of seven alternative electrical generation technologies. The technologies are wind, photovoltaics, hydroelectric, solar thermal, ocean thermal, solid waste, and cogeneration. To perform a comprehensive analysis of the potential of these technologies to produce power in the United States, Frantzis developed a methodology for comparing the data. One of the most important results of this study is her methods: developing a common basis for comparison of the diverse studies and developing a useful range of production scenarios. This article is a synopsis of her thesis. Some of the studies cited by Frantzis are slightly dated, but the findings are no less significant. Her conclusion, that there is abundant potential in the seven alternative technologies to meet our needs, is important because it helps justify continued research and development of a decentralized electrical system.

The objectives of this chapter are to review the major studies of seven alternative technologies for producing electricity; to summarize and standardize the conclusions of these studies for comparison and aggregate analysis; and to assess the technical potential of alternative technologies for producing electricity. Thus, the fundamental question addressed here could be stated as follows: If we commit ourselves to alternative technologies, how much electricity could they realistically produce in the United States, given existing expertise and environmental conditions?

Several issues intentionally omitted from this chapter are discussed elsewhere in this book. They are:

- the economics of the various technologies as they interact in a diversified system;

- present government and industrial regulations and restrictions;
- political obstacles;
- grid integration and management issues; and
- supply and demand planning for stochastic energy sources.

In forecasting future energy supply, planners have often focused on these issues as constraints to developing alternative technologies. But a consideration of constraints too early in the planning process tends to limit energy options for the future. For instance, energy markets are distorted by external costs such as subsidies, government incentives, and price regulations. Often environmental and health costs are not even considered. Alternative technologies may appear uneconomical, given the present means of evaluating and accounting for the costs of conventional systems, because subsidy structures and the absence of a uniform means of comparison prevent balanced judgment. This lack inhibits a clear assessment of the possible technological potential of the alternatives as well as of their relative economics.

Economic and political issues should be considered when formulating a policy or strategy for meeting clear goals. As Brown points out in chapter 1, *once the production potential and desirability of various alternatives have been established, the question of how to make them economically and politically feasible should then be addressed.* In fact, this is how the commitment to nuclear fission was made. Fission was not economically practical in the marketplace in 1950. Only a massive social commitment made it feasible. As was demonstrated with nuclear power, the viability of new technologies cannot be determined solely on the basis of traditional economic analyses. All alternatives must be evaluated by the same criteria and methods.

This chapter offers a preliminary assessment of gross electricity production potential from the following technologies: wind energy conversion, photovoltaics, hydroelectric conversion, solar thermal electric conversion, ocean thermal energy conversion, solid waste conversion, and cogeneration. The gross electricity production capability of these seven alternatives by the year 2000 is assessed, based on the current state of the art of each technology. The data for each energy source are derived from analyses of major national studies written by respected research institutions, private individuals, and government organizations. The maximum electrical outputs are then calculated under the assumption that there could be a total commitment to research, development, and deployment of each of these alternatives.

Past studies have usually focused on only one technology in isolation, and when they have assessed several technologies they have limited the assessment to a discussion of what would be economically feasible or likely to occur based on forecasted trends. If planners and other energy decision-makers are to determine an optimal energy goal for the United States, it is essential that the technological and resource potential for new alternatives to conventional electricity supply sources be assessed and compiled in one report to provide a comparative

THE POTENTIAL FOR DIVERSITY

and cumulative analysis of the alternatives. This study represents a preliminary attempt at such a comprehensive analysis.

BACKGROUND

Though any proposal to diversify the production of electricity has to be based on an assessment of the maximum technological potential of all the alternatives, there is no completely accurate way to project their possible impact and penetration of the market. Therefore, scenarios must be constructed to show a high, low, and average potential. The high estimate can be achieved only with total government and industrial commitment to the technology, whereas the low estimate represents a more realistic amount in light of existing constraints and attitudes. Then the production rates in the various scenarios for each technology can be added to assess their cumulative impact.

It must be understood that the assessed value of any technology—represented in megawatts (MW) or kilowatts (kW)—is predicated on the efficiency of the system. If the rated capacity per unit is in MW or kW, then the total installed capacity can be calculated by multiplying the number of operating units by this value. The actual energy output, however, is not equal to the total installed capacity. For example, a 100-kW wind-energy conversion system may require wind speeds of 12 mph to produce 100 kW. But wind speeds often fall below 12 mph, thus decreasing the average power output of the system. Downtimes for refueling and maintenance work also reduce the actual output of a system. To determine the total actual energy output of a certain technology, we must multiply the installed capacity by the capacity factor, which is the actual amount of electricity generated during one year divided by the amount that would be generated if the plant operated at maximum capacity. As the total actual energy output is also usually given in kilowatt-hours (kWh), we must multiply kW by 8,760 hours per year.

To aid in conceptualizing the actual amount of electricity represented by the calculated figures, kWh are converted into quads or 10^{15} Btu (British thermal units). For the purpose of comparing the electricity generated from alternative energy sources to that generated by conventional systems, one kWh equals 10,000 Btu, because electricity is converted to the equivalent amount of fuel that is saved since it does not have to be burned at a power plant to supply the same amount of power. A total of 10,000 Btu multiplied by the total actual energy output in kWh gives the total actual energy output in quads.

The United States electric utilities consumed 24.8 quads of energy in 1980. The total actual energy output for nuclear power that year was equivalent to 2.7 quads.[1] If we convert the electricity output from the various alternative technologies into a common measurement—quads—we can compare the technologies. Standard charts are presented at the end of each section to summarize and compare the data.

Although solid waste conversion and cogeneration are not necessarily fueled by renewable resources, they are valuable technologies for this assessment because they serve as conservation measures, saving electricity as an interim strategy. Ultimately fossil-fuel cogeneration systems could be converted to hydrogen and other renewable fuels. Conservation efforts will also reduce the value of waste as a fuel source by reducing waste flows. Additional technologies, such as solar water and space heating and bioconversion, will also have an impact on the electricity supply. However, they are more appropriate for reducing electricity consumption and for producing fuels that are better suited for purposes other than generating electricity.

WIND ENERGY CONVERSION

The technology of wind energy conversion systems (WECS) is rapidly becoming more advanced and accepted. Government agencies, utilities, and industrial organizations are beginning to recognize the potential of wind power for meeting energy needs, since both large and small WECS have already been successfully integrated into electricity grids. Moreover, small WECS (between 1.5 and 8 kW) are being used to help meet the electricity demands of individual homes, while the large systems are more commonly being used in industry and agriculture.

Many studies have been conducted to evaluate the possible contribution of WECS by the year 2000. Although the conclusions have varied widely because of differences in values and assumptions, many experts believe that both large and small WECS will play a major role in supplying electricity. In 1979 the Department of Energy (DOE) conducted a domestic policy review of solar energy which resulted in a technical limit estimate of 3 quads for wind systems.[2] These estimates from DOE's *Annual Report to Congress*,[3] published in January 1980, are still being cited. However, higher estimates have been cited in other reports predicated on different underlying assumptions. For example, Marshal F. Merriam, an associate professor in the Department of Materials Sciences at the University of California at Berkeley, estimated in a recent study that wind energy could supply 12 quads of electricity by the year 2000: 10 quads from WECS 0.5 MW and larger, and 2 quads from small WECS ranging from 1 kW to 50 kW (table 1). Merriam's figures represent an "accelerated case" that gives a "reasonable upper bound."[4] The case was based on such assumptions as greater federal financial incentives for WECS, more federal regulations and restrictions in the marketplace, increased demonstration programs, decreased nuclear power capacity, increased fossil fuel prices, and other highly probable events. Based on state-of-the-art and other studies that have been conducted on wind energy potential by the year 2000, a more conservative estimate seems more realistic.

My assessment is based on five studies: DOE's *Domestic Policy Review of Solar Energy*; the Council on Environmental Quality's 1978 report, *Solar Energy Progress and Promise*; the FEA's *Project Independence Final Task Force Report—Solar Energy*; an unpublished Wesleyan University report, "Possible Impact of WECS and Photovoltaics on U.S. Electrical Consumption in 2000"; and Merriam's "Wind Energy Use in the U.S. to the Year 2000."[5]

TABLE 1. Utilization of Wind Energy in the United States (Energy in quads,[a] installed capacity in MW)

	1985		1990		2000	
	Base Case	Accel. Case	Base Case	Accel. Case	Base Case	Accel. Case
Energy from Wind						
Electric Utility						
Mode 1	0.01	0.02	0.08	0.5	1.0	10
Dispersed[b]						
Mode 2	0.009	0.015	0.05	0.2	0.4	2
Installed Capacity						
Electric Utility						
Mode 1	330	650	2600	16,000	33,000	330,000
Dispersed[b]						
Mode	290	490	1600	6,500	13,000	65,000
Number of Machines						
Electric Utility						
Mode 1						
0.5 MW	170	380	400	4,000	1,200	40,000
1.0 MW	200	400	2,000	12,000	20,000	210,000
2.0 MW	20	30	200	1,000	6,000	50,000
Total	390	810	2,600	17,000	27,000	300,000
Dispersed[b]						
Mode 2						
1 kW	10,000	20,000	50,000	100,000	200,000	1,000,000
5 kW	10,000	20,000	50,000	100,000	200,000	1,000,000
10 kW	10,000	20,000	43,000	140,000	400,000	1,400,000
20 kW	5,000	6,000	30,000	100,000	140,000	1,000,000
50 kW	600	1,000	6,000	50,000	100,000	500,000
Total	36,000	67,000	180,000	490,000	1,000,000	4,900,000

a. 1 quad = 10^{15} Btu. Wind machine outputs are electrical, and impact the national energy situation through saving oil in thermal power stations. Thus, wind machine outputs have been converted to quads by setting 1 kWh = 10,000 Btu.
b. "Dispersed" is residences + industry + agriculture + miscellaneous, of which residences is about two-thirds of the total.
Note: All entries are rounded to two significant figures. "Energy" entries are related to "capacity" entries through an arbitrarily assumed plant capacity factor of 0.35.
Source: Marshal F. Merriam, "Wind energy use in the U.S. to the year 2000," Wind Report, August 1978.

For large WECS, most of the current DOE-funded research and development has been aimed at MW-size units. Thus, a rated capacity per unit of 1 MW can be assumed for large WECS. In comparison with the Merriam and Project Independence reports, a conservative cumulative WECS production and deployment figure for the year 2000 ranges from 40×10^3 to 90×10^3 units. The total installed capacity can be calculated as follows:

$$(1 \text{ MW}) (40 \times 10^3 \text{ units}) = 40 \times 10^3 \text{ MW [low]}$$
$$(1 \text{ MW}) (90 \times 10^3 \text{ units}) = 90 \times 10^3 \text{ MW [high]}$$

Taking into account a capacity factor of 25% (which is very conservative compared with the Merriam and Project Independence reports), we find the total actual energy output for all large WECS to be:

$$(40 \times 10^3 \text{ MW}) (.25) = 10 \times 10^3 \text{ MW or } 10 \times 10^6 \text{ kW}$$
$$= 88 \times 10^9 \text{ kWh}$$
$$= 0.88 \text{ quad [low]}$$
$$(90 \times 10^3 \text{ MW}) (.25) = 22.5 \times 10^3 \text{ MW or } 22.5 \times 10^6 \text{ kW}$$
$$= 197 \times 10^9 \text{ kWh}$$
$$= 2 \text{ quads [high]}$$

For intermediate WECS, the most common rated capacity per unit is 200 kW. If we assume a production and deployment range of 250×10^3 to 800×10^3 units by the year 2000, then the total installed capacity is:

$$200 \text{ kW} \times 250 \times 10^3 \text{ units} = 50 \times 10^3 \text{ MW [low]}$$

or

$$200 \text{ kW} \times 800 \times 10^3 \text{ units} = 160 \times 10^3 \text{ MW [high]}$$

The total actual energy output for all intermediate WECS, assuming a capacity factor of 25%, is calculated as follows:

$$(50 \times 10^3 \text{ MW}) (.25) = 12.5 \times 10^3 \text{ MW or } 12.5 \times 10^6 \text{ kW}$$
$$= 110 \times 10^9 \text{ kWh}$$
$$= 1.1 \text{ quads [low]}$$

or

$$(160 \times 10^3 \text{ MW}) (.25) = 40 \times 10^3 \text{ MW or } 40 \times 10^6 \text{ kW}$$
$$= 350 \times 10^9 \text{ kWh}$$
$$= 3.5 \text{ quads [high]}$$

For small WECS, if we assume an ideally rated unit of 8 kW for a home and a production and deployment range of $1,000 \times 10^3$ to $1,500 \times 10^3$ units, then the total installed capacity is:

$$(8 \text{ kW}) (1,000 \times 10^3 \text{ units}) = 8 \times 10^3 \text{ MW [low]}$$

or

$$(8 \text{ kW}) (1,500 \times 10^3 \text{ units}) = 12 \times 10^3 \text{ MW [high]}$$

The total actual energy output for all small WECS is:

$$(8 \times 10^3 \text{ MW}) (.25) = 2 \times 10^3 \text{ MW or } 2 \times 10^6 \text{ kW}$$
$$= 18 \times 10^9 \text{ kWh}$$
$$= 0.18 \text{ quad [low]}$$
$$(12 \times 10^3 \text{ MW}) (.25) = 3 \times 10^3 \text{ MW or } 3 \times 10^6 \text{ kW}$$
$$= 26 \times 10^9 \text{ kWh}$$
$$= 0.26 \text{ quad [high]}$$

TABLE 2. Wind Energy Conversion Potential by the Year 2000

	Rated Capacity /Unit	No. of Units	Total Installed Capacity (MW×10³)	Total Actual Energy Output[a] (kWh×10⁹)	Total Actual Energy Output (quads)[b]	Percent 1980 U.S. Electricity Demand[c]	Percent 1980 U.S. Nuclear Equiv.[d]
Large							
Low	1 MW	40×10³	40	88	0.88	3.6	33
High	1 MW	90×10³	90	197	2	8.2	74
Intermediate							
Low	200 kW	250×10³	50	110	1.1	4.5	41
High	200 kW	800×10³	160	350	3.5	14	130
Small							
Low	8 kW	1000×10³	8	18	0.18	0.7	6.7
High	8 kW	1500×10³	12	26	0.26	1.1	9.6
Total							
Low		1290×10³	98	216	2.2	9	82
High		2390×10³	262	573	5.8	24	215
Average		1840×10³	180	395	4	16	148

a. Assuming a capacity factor of 25%.
b. 1 kWh=10,000 Btu. Electricity is converted to equivalent fuel that is saved by not having to be burned at a power plant to supply the same amount of power.
c. 1980 U.S. electricity demand: 24.8 quads.
d. 1980 U.S. nuclear generation: 2.7 quads.

Thus, the totals for all WECS are:

Low 2.16 quads
High 5.73 quads
Average 4 quads = 16% of the 1980 U.S. electricity demand and 148% of the 1980 equivalent of nuclear power generation

Thus, it is evident that wind could contribute a significant amount of electricity by the year 2000 (table 2). Many sites in the United States, as noted in a 1981 Arthur D. Little, Inc., report, have average wind velocities of at least 12 mph, making them appropriate sites for WECS of all sizes.[6] Some of the most promising areas in the United States for large-scale development of wind systems are in the mountainous regions, the Central and Great Plains, the northeast coast, the coast of Alaska, Montana, and Wyoming. In Wyoming, for example, the average annual wind speed in many areas is greater than 17.9 mph at an elevation of 50 m. Resources are available to achieve the calculated 4 quads of electricity, but a serious commitment to WECS is needed before limiting factors involving aesthetics, costs, and environmental concerns can be overcome.

PHOTOVOLTAICS

The Council on Environmental Quality has calculated that by the turn of the century the installed capacity of solar cells could reach 75,000 MW.[7] This amount is equivalent to an estimated 2 to 8 quads of electricity. Numerous documents, however, describe more advanced improvements in solar cell devices that have not even been considered in the calculations. One such improvement is the use of gallium arsenide (GaAs) cells rather than silicon cells. Because of their higher efficiency and ability to withstand intensified insolation, GaAs cells could significantly increase the energy output and potential applications for solar cells. Today, commonly used silicon cells generate approximately 150 W/m^2 or 0.15 kW/m^2 of cell area.[8] By experimenting with GaAs cells, researchers at the Varian Corporation succeeded in concentrating the sun's rays 1,735 times. By coupling the concentrator with the collector, they achieved an output of 0.24 MW/m^2 of cell area at a 19% efficiency.[9] This achievement may be useful in promoting larger-scale applications for solar cells that have not yet been considered. A feasible alternative would be to place several arrays at one location within a town. Coupled with a powerful concentrator, GaAs arrays could help supply electricity to meet local needs.

For large multi-array units of GaAs solar cells, we could assume one-half the achieved 240 kW/m^2 of cell area. An array with 20 m^2 of cell area will thus have a rated capacity per unit of 120 kW/m^2 × 20 m^2 = 2.4 MW. If we assume a production and deployment range of 1×10^4 to 5×10^4 units by the year 2000, the total installed capacity would be:[10]

$$(2.4 \text{ MW}) (1 \times 10^4 \text{ units}) = 24 \times 10^3 \text{ MW [low]}$$

or

$$(2.4 \text{ MW}) (5 \times 10^4 \text{ units}) = 120 \times 10^3 \text{ MW [high]}$$

THE POTENTIAL FOR DIVERSITY

Taking into consideration the fact that the sun shines only an average six hours a day and that there are occasional extremely cloudy days, we must incorporate a capacity factor of 20% to calculate the total actual energy output. Thus,

$$(24 \times 10^3 \text{ MW}) (.20) = 4.8 \times 10^3 \text{ MW or } 4.8 \times 10^6 \text{ kW}$$
$$= 42 \times 10^9 \text{ kWh}$$
$$= 0.42 \text{ quad [low]}$$

or

$$(120 \times 10^3 \text{ MW}) (.20) = 24 \times 10^3 \text{ MW or } 24 \times 10^6 \text{ kW}$$
$$= 210 \text{ kWh}$$
$$= 2.1 \text{ quads [high]}$$

For intermediate multi-array and single-array use of GaAs cells, we may assume the use of a concentrator that magnifies the light intensity 12-fold versus 1,735-fold—the level attained by the Varian Corporation; the resulting rated output is 1.7 kW/m². The rated capacity per unit is 1.7 kW/m² × 20 m² = 34 kW. If we assume a production and deployment range of 199×10^4 to 295×10^4 units by the year 2000, the total installed capacity would be:

$$(34 \text{ kW}) (199 \times 10^4 \text{ units}) = 68 \times 10^3 \text{ MW [low]}$$

or

$$(34 \text{ kW}) (295 \times 10^4 \text{ units}) = 100 \times 10^3 \text{ MW [high]}$$

The total actual energy output for all intermediate solar cells, if we assume a capacity factor of 20%, can be calculated as follows:

$$(68 \times 10^3 \text{ MW}) (.20) = 13.6 \times 10^3 \text{ MW or } 13.6 \times 10^6 \text{ kW}$$
$$= 119 \times 10^9 \text{ kWh}$$
$$= 1.2 \text{ quads [low]}$$

or

$$(100 \times 10^3 \text{ MW}) (.20) = 20 \times 10^3 \text{ MW or } 20 \times 10^6 \text{ kW}$$
$$= 175 \times 10^9 \text{ kWh}$$
$$= 1.8 \text{ quads [high]}$$

Unlike GaAs cells, silicon cells would be suitable for new and retrofitted buildings. By the year 2000 at least 40 million new buildings will be constructed.[11] An array with 20 m² of cell area could fit easily on the roofs of most buildings. A packing capacity of 85%, for example, would require only 23 m² of roof area for the panel of solar cells. For small-scale silicon cells, the calculated output is 0.15 kW/m² and the rated capacity per unit is 3 kW. If we assume a production and deployment range of 400×10^4 to 700×10^4 units by the year 2000, then the total installed capacity would be:

$$(3 \text{ kW}) (400 \times 10^4 \text{ units}) = 12 \times 10^3 \text{ MW [low]}$$

or

$$(3 \text{ kW}) (700 \times 10^4 \text{ units}) = 21 \times 10^3 \text{ MW [high]}$$

The total actual energy output, again if we assume a capacity factor of 20%, would be:

$$(12 \times 10^3 \text{ MW}) (.20) = 2.4 \times 10^3 \text{ MW or } 2.4 \times 10^6 \text{ kW}$$
$$= 21 \times 10^9 \text{ kWh}$$
$$= .21 \text{ quad [low]}$$

or

$$(21 \times 10^3 \text{ MW}) (.20) = 4.2 \times 10^3 \text{ MW or } 4.2 \times 10^6 \text{ kW}$$
$$= 37 \times 10^9 \text{ kWh}$$
$$= .37 \text{ quad [high]}$$

The total for all solar cells would be:

Low   1.8 quads
High   4.3 quads
Average 3 quads = 12% of the 1980 U.S. electricity demand and 111% of the 1980 equivalent of nuclear power generation.

Many major breakthroughs in solar cell technology have already been made. Mass production of these arrays could easily reach the number of arrays assumed in the above calculations. As the technology improves and the efficiency of solar cells increases, more electricity will be generated from the same number of cells. It should be emphasized that these figures are all based on the present state of the art. By the year 2000 the technology should have improved substantially, thus making the above calculations seem very conservative.

The energy output figures (table 3) may seem unrealistic, but we must remember that these figures represent what *could*—not necessarily what *should*—occur by the year 2000, if an effort were made to achieve these goals. The technology is here; government, industries, and citizens should be made aware of the potential and strive to meet electricity demands with photovoltaics.

HYDROELECTRIC CONVERSION

The possible advantages of (1) renovating old hydroelectric sites; (2) converting existing dams into hydroelectric facilities; and (3) improving the efficiency of hydroelectric plants currently in operation have been the focal points of recent major studies. The most detailed and thorough study was conducted by the U.S. Army Corps of Engineers in 1977. Based primarily on the results of this study, DOE has already taken action to continue research and feasibility studies, and to develop plans for implementation.

The Army Corps of Engineers' study estimates that there is a potential for generating 54,600 MW, or 159.3 billion kWh/year, of power at existing large and small dams in the United States.[12] In comparison, a study done by Energy Research and Applications in 1976 proposed that a minimum capacity of 14,000 MW can be achieved by retrofitting small dam sites (less than 5 MW). An addi-

TABLE 3. Photovoltaic Potential by the Year 2000

	Rated Capacity /Unit[a]	No. of Units	Total Installed Capacity (MW × 10³)	Total Actual Energy Output[b] (kWh × 10⁹)	Total Actual Energy Output (quads)[c]	Percent 1980 U.S. Electricity Demand[d]	Percent 1980 U.S. Nuclear Equiv.[e]
Large (GaAs)							
Low	2.4 MW	1 × 10⁴	24	42	.42	1.8	16
High	2.4 MW	5 × 10⁴	120	210	2.1	8.6	78
Intermediate							
Low	34 kW	199 × 10⁴	68	119	1.2	4.9	44
High	34 kW	295 × 10⁴	100	175	1.8	7.4	67
Small (silicon)							
Low	3 kW	400 × 10⁴	12	21	.21	.86	7.8
High	3 kW	700 × 10⁴	21	37	.37	1.5	14
Total							
Low		600 × 10⁴	104	182	1.8	7.4	67
High		1000 × 10⁴	241	422	4.3	18	159
Average		800 × 10⁴	173	302	3	12	111

a. Array size: 420 ft² or 20 m²; also includes conversion efficiency.
b. Assuming a capacity factor of 20%.
c. 1 kWh = 10,000 Btu. Electricity is converted to equivalent fuel that is saved by not having to be burned at a power plant to supply the same amount of power.
d. 1980 U.S. electricity demand: 24.8 quads.
e. 1980 U.S. nuclear generation: 2.7 quads.

tional 14,000 MW can be generated from existing dam sites with potentials greater than 5 MW.[13]

The discrepancies about the potential of hydroelectric power result from each report's being predicated on different assumptions. The Army Corps report, for example, takes into consideration the installation of more efficient and greater numbers of turbines and generators at existing hydroelectric sites. Many other studies just account for the implementation of facilities at all nonhydropower dams; thus the figures in the Army Corps study are slightly higher than those in some of the other reports.

Of all the reports reviewed, the largest potential for hydroelectric power by the year 2000 was estimated by the Council on Environmental Quality (CEQ); it was between 4 and 6 quads. In calculating estimates for the year 2000, however, most of the data were derived from the Army Corps study, which had not been privy to the projections derived from their research.

Hydroelectric power from existing dams can supply a significant amount of energy by the year 2000 without serious environmental impacts. When the Army Corps study was conducted in 1977, 11% of the demand for electricity in the United States, or 271 billion kWh (2.7 quads), was generated by hydroelectric facilities. An additional .4 quad of power has since been obtained from these facilities. The total actual energy output in 1980 was 3.13 quads—equivalent to 12.6% of the demand for electricity in the United States in 1980.[14]

There are other potential sites for hydroelectric facilities in addition to those already developed or now under construction. As discussed in the Army Corps study, there is a potential for 54.2 billion kWh if some existing dams are rehabilitated and expanded. For the potential rehabilitation and expansion of existing dams, we can assume a high production and deployment of 100% of the Army Corps estimate and a low production and deployment of 50% by the year 2000. That is:

$$(54.2 \times 10^9 \text{ kWh}) (.50) = 27.1 \times 10^9 \text{ kWh, or } 0.27 \text{ quad [low]}$$

or

$$(54.2 \times 10^9 \text{ kWh}) \text{ or } 0.54 \text{ quad [high]}$$

For the potential at existing nonhydroelectric dams, again we can assume 100% of the Army Corps figures for a high estimate and 50% of the Army Corps figures for a low estimate. That is:

$$(105.1 \times 10^9 \text{ kWh}) (.50) = 52.6 \times 10^9 \text{ kWh or } 0.53 \text{ quad [low]}$$

or

$$(105.1 \times 10^9 \text{ kWh}) \text{ or } 1.0 \text{ quad [high]}$$

The total energy potential for undeveloped dam sites is thus:

Low   0.80 quad
High   1.6 quads

Average   1.2 quads = 5% of the 1980 U.S. electricity demand and 44% of the 1980 equivalent of nuclear power generation

The total energy potential for both developed and undeveloped sites by the year 2000 is:

Low   3.7 quads
High   4.5 quads
Average   4.1 quads = 17% of the 1980 U.S. electricity demand and 152% of the 1980 equivalent of nuclear power generation

The contribution of hydroelectric power by the year 2000 can be significant, as the data above and table 4 illustrate. A strong social commitment must be made before proper development can occur. Only 1,400 of the 50,000 existing small dams in the United States have been developed for hydroelectric power. Many of these dams are being used for flood control, irrigation, or river navigation or recreation and would only have to be retrofitted with turbines and other minor hydroelectric parts: "If only 10% of our 50,000 small dams were even partially developed, we could save the energy equivalent of 180 million barrels of oil every year."[15]

The greatest potential for significant development of hydroelectric power is in New England, the upper and lower Mississippi River areas, Alaska, the Middle Atlantic states and the Pacific Northwest. New England alone has more than 9,000 dams without hydroelectric systems, and more than 228 abandoned small hydroelectric facilities.[16] A recent study conducted by the New England River Basin Commission concluded that there is potential for the development of 1,000 MW at 1,750 unused small dam sites in the area.[17] Detailed studies of hydroelectric potential are being conducted throughout the United States. The water power is available, but measures must be taken to tap this source of energy.

SOLAR THERMAL ENERGY CONVERSION

Solar thermal energy conversion (STEC) is the process of collecting solar heat and converting it into electricity by a thermomechanical process. Primary emphasis in the United States is on STEC devices that concentrate the solar energy to build high temperatures to produce steam, which in turn powers electricity-generating equipment. The study summarized here discusses only the central receiver or power tower concept, those facilities being funded and tested by DOE, and concentrates on those systems considered for use in the southwestern United States. The Southwest has the greatest potential for STEC systems because it receives twice as much direct insolation as the Midwest, and roughly two-thirds more insolation than the Northeast and Northwest. Unlike flat-plate collectors, which utilize diffuse sunlight on hazy days, STEC systems require bright, clear days for design condition operation. The high incidence of solar radiation, combined with the vast unused desert lands of the Southwest, yield optimal conditions for STEC development.

TABLE 4. Hydroelectric Potential by the Year 2000

		Total Installed Capacity (MW×10³)	Total Actual Energy Output (kWh×10⁹)	Total Actual Energy Output (quads)[a]	Percent 1980 U.S. Electricity Demand[b]	Percent 1980 U.S. Nuclear Equiv.[c]
Developed		57	271	2.7	11	100
Under Construction		8.2	16.8	.17	.70	6.3
Total Installed		65.2	287.8	2.9	12	107
Potential Rehabilitation and Expansion of Existing Dams	Low	10.5	27.1	.27	1.1	10
	High	21	54.2	.54	2.2	20
Potential at Existing Non-Hydroelectric Dams	Low	16.8	52.6	.53	2.2	20
	High	33.6	105.1	1.0	4.0	37
Total Potential (undeveloped)	Low	27.3	79.7	.80	3.3	30
	High	54.6	159.3	1.6	6.6	60
	Average	40.9	119.5	1.2	5.0	44
Total (developed and undeveloped)	Low	92.5	367.5	3.7	15	137
	High	119.8	447.1	4.5	18	167
	Average	106.2	407.3	4.1	17	152

a. 1 kWh = 10,000 Btu. Electricity is converted to equivalent fuel that is saved by not having to be burned at a power plant to supply the same amount of power.
b. 1980 U.S. electricity demand: 24.8 quads.
c. 1980 U.S. nuclear generation: 2.7 quads.

THE POTENTIAL FOR DIVERSITY

The use of 10,000 square miles or about 10% of the California-Arizona desert for electric power generation could provide an installed capacity of about 1 million MW, or more than twice the present U.S. generating capacity.[18]

In 1970 electricity demands in the Southwest required intermediate and peak generation capacities of 19,049 MW and 8,960 MW, respectively. Although baseload generation capacity was only 20,000 MW, conventional power plants averaged a capacity factor of 80% in comparison with a capacity factor of 42% generally associated with intermediate load systems (table 5). STEC systems without storage, therefore, cannot adequately supply baseload needs because of problems in attaining a reliability of 80% due to inconsistent weather patterns and lack of daily sunshine. Baseload electricity demands can be met only if (1) storage is used; (2) the STEC system is used to produce hydrogen that can be stored or fed into a modified natural gas pipeline; or (3) fossil fuels are used to provide backup.

In assessing a plausible deployment for STEC systems by the year 2000, however, it would be more advantageous and realistic to consider STEC systems just for intermediate or peak-load supply.[19] By the year 2000, Project Independence forecasts an intermediate load of 675×10^9 kWh and a peak load of 10×10^9 kWh. If present trends of electricity consumption do not continue, and the limitations of forecasting methods are recognized, then we could assume that these calculated values would be much lower.

Two reports—*Solar Energy Progress and Promise*, by the Council on Environmental Quality, and the *Project Independence Final Task Force Report—Solar Energy*, by the Federal Energy Administration—estimate a potential of at least 40,000 MW for large-scale STEC by the year 2000. The former report claims that 0–2 quads by the year 2000 and eventually 20–30% of total United States electricity needs could be generated by STEC power.[20] There are, however, three different scales of operation: (1) large STEC systems ranging from 20 to 200 MW, which are appropriate for utility-scale power generation requirements; (2) intermediate STEC systems ranging from 1 to 10 MW, which can be used to supply energy to small community systems and total energy systems; and (3) small STEC systems ranging from 200 to 600 kW, which can be used for relatively small, on-site applications. Most of the research and development will be focused on large STEC systems (around 100 MW); thus we can assume an average rated capacity per unit of 100 MW for a large STEC system. A feasible production and deployment range by the year 2000, as discussed in the Project Independence report, might be 40,000 MW with a "base case" and 80,000 MW with an "accelerated case." We can therefore assume a conservative production and deployment range of 2×10^2 to 4×10^2 units by the year 2000. The total installed capacity would thus be:

$$(100 \text{ MW}) (2 \times 10^2 \text{ units}) = 20 \times 10^3 \text{ MW [low]}$$

or

$$(100 \text{ MW}) (4 \times 10^2 \text{ units}) = 40 \times 10^3 \text{ MW [high]}.$$

TABLE 5. Generation Capacity/Energy Demand Estimates for Southwestern Region

Year	Peak Demand (MWe)	Generation Capacity[a] (MWe)				Energy Demand[b] (kWh × 10^9)			
		Total	Base	Intermed.	Peaking	Total	Base	Intermed.	Peaking
1970	40,000	48,000	20,000	19,040	8,960	211	140	70	1
1980	89,000	106,800	44,500	42,360	19,940	470	312	156	2
1990	195,000	234,000	97,500	92,820	43,680	1,030	683	342	5
2000	385,000	462,000	192,500	183,260	86,240	2,034	1,349	675	10

a. Generation Capacity = 1.2 × peak demand (20% margin)
   Base-Load Capacity = 0.500 × peak demand (capacity factor = 0.80)
   Intermed. Capacity = 0.476 × peak demand (capacity factor = 0.42)
   Peaking Capacity = 0.224 × peak demand (capacity factor = 0.07)
b. Total Energy = 0.6 × peak demand × 8,760
   Base-Load Energy = 0.663 × total energy
   Intermed. Energy = 0.332 × total energy
   Peaking Energy = 0.005 × total energy

Source: Federal Energy Administration, Project Independence Final Task Force Report—Solar Energy, 1974.

THE POTENTIAL FOR DIVERSITY

Taking into account a capacity factor of 40%, which is very conservative in comparison with the 55% capacity factor used in the Project Independence report,[21] we can calculate the total actual energy output for all large units as follows:

$$(20 \times 10^3 \text{ MW}) (.40) = 8 \times 10^3 \text{ MW or } 8 \times 10^6 \text{ kW}$$
$$= 70 \times 10^9 \text{ kWh}$$
$$= .7 \text{ quad [low]}$$

or

$$(40 \times 10^3 \text{ MW}) (.40) = 16 \times 10^3 \text{ MW or } 16 \times 10^6 \text{ kW}$$
$$= 140 \times 10^9 \text{ kWh}$$
$$= 1.4 \text{ quads [high]}$$

For intermediate STEC systems ranging from 1 to 10 MW, if we assume an average rated capacity per unit of 5 MW and a production and deployment range of 40×10^2 to 90×10^2 units by the year 2000, then the total installed capacity would be:

$$(5 \text{ MW}) (40 \times 10^2 \text{ units}) = 20 \times 10^3 \text{ MW [low]}$$

or

$$(5 \text{ MW}) (90 \times 10^2 \text{ units}) = 45 \times 10^3 \text{ MW [high]}$$

The total actual energy output for all intermediate STEC systems, if we assume a capacity factor of 40%, can be calculated as follows:

$$(20 \times 10^3 \text{ MW}) (.40) = 8 \times 10^3 \text{ MW or } 8 \times 10^6 \text{ kW}$$
$$= 70 \times 10^9 \text{ kWh}$$
$$= .7 \text{ quad [low]}$$

or

$$(45 \times 10^3 \text{ MW}) (.40) = 18 \times 10^3 \text{ MW or } 18 \times 10^6 \text{ kW}$$
$$= 158 \times 10^9 \text{ kWh}$$
$$= 1.6 \text{ quads [high]}$$

For small STEC systems we can assume an average rated capacity per unit of 400 kW—an ideal size for the electricity needs of a small community. We can also assume a production and deployment range of 250×10^2 to 400×10^2 units by the year 2000. Thus, the total installed capacity would be:

$$(400 \text{ kW}) \times (250 \times 10^2 \text{ units}) = 10 \times 10^3 \text{ MW [low]}$$

or

$$(400 \text{ kW}) \times (400 \times 10^2 \text{ units}) = 16 \times 10^3 \text{ MW [high]}$$

The total actual energy output for all small STEC systems would be:

$$(10 \times 10^3 \text{ MW}) (.40) = 4 \times 10^3 \text{ MW or } 4 \times 10^6 \text{ kW}$$
$$= 35 \times 10^9 \text{ kWh}$$
$$= .35 \text{ quad [low]}$$

or

$$(16 \times 10^3 \text{ MW}) (.40) = 6.4 \times 10^3 \text{ MW or } 6.4 \times 10^6 \text{ kW}$$
$$= 56 \times 10^9 \text{ kWh}$$
$$= .56 \text{ quad [high]}$$

The total for all STEC systems would be:

Low   1.8 quads
High   3.6 quads
Average   2.7 quads = 11% of the 1980 U.S. electricity demand and 100% of the 1980 equivalent of nuclear power generation

The capacity attributable to STEC systems by the year 2000 could help satisfy intermediate or peak loads in the Southwest, with the production and deployment rate presented above (table 6). Both land and resources are available for further development of STEC systems. Exploiting the full energy potential of STEC systems in the deserts of the Southwest, however, would be unnecessary and would place too much dependence on one power source. Massive use of this technology would require storage for export to other regions, which in turn would reduce system efficiency and increase costs. Coincidence between daily demand and maximum daily insolation, therefore, makes STEC systems an appropriate technology to help satisfy the intermediate or peak-load demands of the Southwest by the year 2000.

OCEAN THERMAL ENERGY CONVERSION

Ocean thermal energy conversion (OTEC) utilizes the thermal gradient in the ocean waters to produce electricity. Although OTEC is one of the only solar technologies that has never been tested on a commercial scale (i.e., MW size), we may be optimistic about the outlook for the year 2000 if development programs are successful. The best sites for OTEC systems in the United States are in the Gulf of Mexico, the Caribbean, the Hawaiian Islands, and the Gulf Stream waters off the southeastern coast (figure 1). In the Gulf of Mexico alone, DOE has estimated a potential of 200,000–600,000 MW, using 500–1,500 OTEC systems to produce 1.4 trillion to 4 trillion kWh (15 to 40 quads annually).[22] The Project Independence report calculated a low of 65,000 MW (4.5 quads, assuming a capacity factor of 80%) and a high of 260,000 MW (18.2 quads) for all United States OTEC systems by the year 2000.[23] In comparison, the Committee on Nuclear and Alternative Energy Systems (CONAES) estimated 1.6 quads, the Council on Environmental Quality (CEQ) estimated 1 to 3 quads, and the DOE in its *Annual Report to Congress* (1980) estimated 1 quad by the year 2000.[24]

TABLE 6. Solar Thermal Electric Potential by the Year 2000

	Rated Capacity /Unit	No. of Units	Total Installed Capacity (MW × 10³)	Total Actual Energy Output[a] (kWh × 10⁹)	Total Actual Energy Output (quads)[b]	Percent 1980 U.S. Electricity Demand[c]	Percent 1980 U.S. Nuclear Equiv.[d]
Large							
Low	100 MW	2×10^2	20	70	.7	2.9	26
High	100 MW	4×10^2	40	160	1.4	5.7	52
Intermediate							
Low	5 MW	40×10^2	20	70	.7	2.9	26
High	5 MW	90×10^2	45	158	1.6	6.6	59
Small							
Low	400 kW	250×10^2	10	35	.35	1.4	13
High	400 kW	400×10^2	16	56	.55	2.3	20
Total							
Low		292×10^2	50	175	1.8	7.4	67
High		494×10^2	101	374	3.6	15	133
Average		393×10^2	76	275	2.7	11	100

a. Assuming a capacity factor of 40%.
b. 1 kWh = 10,000 Btu. Electricity is converted to equivalent fuel that is saved by not having to be burned at a power plant to supply the same amount of power.
c. 1980 U.S. electricity demand: 24.8 quads.
d. 1980 U.S. nuclear generation: 2.7 quads.

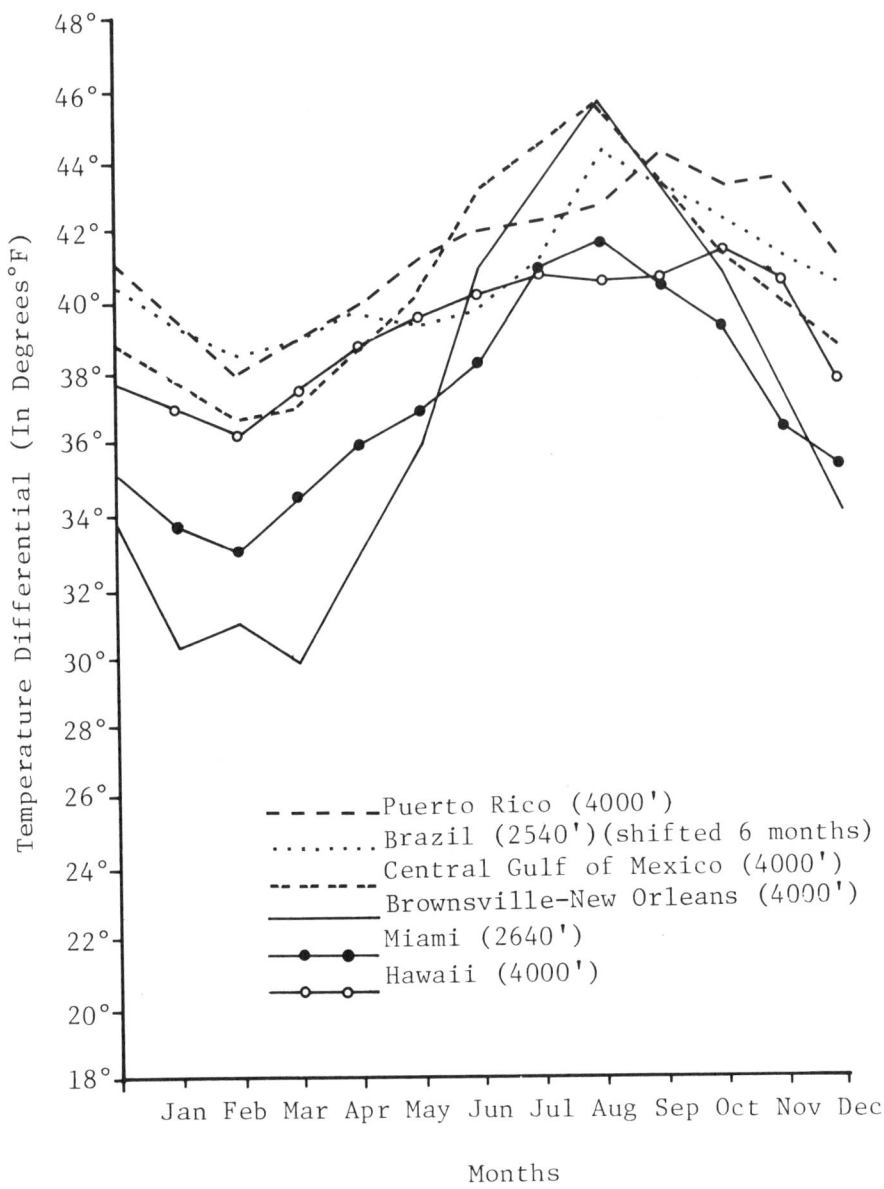

FIGURE 1. Monthly Thermal Resources Comparison for Selected OTEC Sites. Source: Committee on Science and Technology, *Energy from the Ocean*, April 1978.

As these figures indicate, assessments for the electricity potential from OTEC by the year 2000 vary considerably. The calculations in this report have therefore been derived from an evaluation of the assumptions and calculations of each study. The average rated capacity per unit is expected to be 400 MW, if we assume a production and deployment level of from 50 to 125 units by the year 2000.

$$(50 \times 10^3 \text{ MW}) (.80) = 40 \times 10^3 \text{ MW or } 40 \times 10^6 \text{ kW}$$
$$= 350 \times 10^9 \text{ kWh}$$
$$= 3.5 \text{ quads [high]}$$

The total for all OTEC systems would be:

Low   1.4 quads
High 3.5 quads
Average 2.5 quads = 10% of the 1980 U.S. electricity demand and 93% of the 1980 U.S. equivalent of nuclear power generation.

The nature of OTEC systems is such that, in principle, they are well suited to baseload power generation, since the thermal energy is available without diurnal variation or interruption. The technology appears promising for the near future, as government program objectives for OTEC systems call for OTEC facilities to produce electricity for grid integration and for on-site production of energy-intensive products.

SOLID WASTE CONVERSION

Americans generate more than 2 billion tons of organic waste each year. Roughly 160 million tons are from municipal solid waste. Systems are currently available that convert solid waste into useful fuels for electricity production, but the technology has not been utilized to its full potential. Resource recovery will become very economical as land prices and the costs of abiding by environmental regulations and transporting wastes increase. Each year 12 million tons of steel, 1 million tons of aluminum, and 200,000 tons of copper are wasted; only 6 to 7% of solid waste is recycled. The Environmental Protection Agency (EPA) estimates that a "nationwide source separation effort could recycle approximately 25% of the nation's total gross discards."[25]

Various studies have estimated that by 1985 approximately 200 million tons of waste will be generated. An EPA report (*Fourth Report to Congress: Resource Recovery and Waste Reduction*) claims that the greatest potential for energy recovery from waste is in urban areas, where 70% of the nation's waste is generated. The capacity factor of waste recovery in urban areas is 80%.

We can assume that the amount of waste that will be generated by the year 2000 will range from 200 to 300 million tons. A low of 200 million tons, which is the same amount of waste estimated for 1985, can be assumed to be generated if strict conservation measures are enforced, and a high of 300 million tons can be assumed if present trends of consumption and waste generation continue with only some conservation strategies. It can also be assumed that the collection and recovery of solid waste will occur in urban areas. The maximum amount that is recoverable would thus be 56% (80% × 70% = 56%).

One ton of solid waste generates approximately 700 kWh (table 7). If a low of 200 million tons is generated by the year 2000, about 0.8 quad of energy can be recovered:

$$(200 \times 10^6 \text{ tons}) (.56) = 112 \times 10^6 \text{ tons}$$

or

$$(112 \times 10^6 \text{ tons}) (700 \text{ kWh/ton}) = .78 \times 10^{11} \text{ kWh}$$
$$= .8 \text{ quad [low]}$$

If a high of 300 million tons is generated by the year 2000, about 1.5 quads of energy could be recovered:

$$(300 \times 10^6 \text{ tons}) (.56) = 168 \times 10^6 \text{ tons}$$

or

$$(168 \times 10^6 \text{ tons}) (700 \text{ kWh/ton}) = 1.2 \times 10^{11} \text{ kWh}$$
$$= 1.2 \text{ quads [high]}$$

The total for all solid waste conversion would be:

Low   0.8 quad
High   1.2 quads
Average 1.0 quads = 4.0% of the 1980 U.S. electricity demand and 37% of the 1980 equivalent of nuclear power generation.

Solid waste conversion systems should be constructed close to the major areas of waste generation to promote efficiency. A city should evaluate its population density, industrial waste, and transportation system before constructing a facility in the area. Some locations may not even be able to supply adequate amounts of waste to make the system cost-effective.

Many cities are in great need of recycling and fuel-recovery operations. Recycling measures and more durable goods could reduce the waste of glass by 40%, rubber tires by 40%, aluminum by 30%, ferrous metals by 15%, and other materials, such as paper, by 10%.[26] Conservation measures should be promoted just as much as solid waste conversion. Both would reduce our demand on di-

TABLE 7. Energy Recovered from 1 Ton of Municipal Solid Waste

Product	Output
Electricity	700 kWh
Steam	6000 lb
Gas	20,000 ft^3
Oil	1 bbl
RDF	0.7 ton

Source: U.S. Environmental Protection Agency, *Resource Recovery Technology, An Implementation Seminar Workbook*, 1977.

THE POTENTIAL FOR DIVERSITY

minishing fossil fuel reserves without necessitating a dramatic change in America's high standard of living.

COGENERATION

Industrial operations require approximately 40% of the fuel consumed in the United States, but more than one-third of this energy is wasted. Cogeneration could yield tremendous savings by the year 2000 by utilizing energy that is being wasted in another consuming process. "Cogeneration and more efficient use of electricity could together reduce our use of electricity by a third and our centralization generation by 60%."[27] Utility companies are currently having difficulty meeting electricity demands at low prices. It has finally become economical for people to consider alternatives like cogeneration, which have the potential to become major sources for electricity production.

The three most comprehensive studies estimating the potential contribution of cogeneration systems by the year 2000 are *Industrial Cogeneration*, by Robert H. Williams; *Cogeneration: An Assessment of Commercial Readiness*, published by DOE; and *The Potential for Cogeneration Development in Six Major Industries by 1985*, prepared by Resource Planning Associates, Inc. (RPA), for DOE. The maximum *conceivable* potential by the year 2000 is Williams's estimate of 1,000 gigawatts (GW), or 48 quads, assuming a conservative capacity factor of 55%. This amount could be obtained "by associating all process steam with Diesel cogeneration units."[28] Williams's estimate is not likely to be achieved, however, given the present emphasis on coal usage and the diminishing reserves of natural gas and distillate oils. A more conservative estimate in his study assumes that the potential steam load in industrial facilities by 2000 will be 17.9 quads, a sum estimated on the assumption that fuel conservation strategies will be incorporated to reduce the demand for electricity. If 17.9 quads are used in industrial facilities, Williams's report claims, 208 GW could reasonably be generated by cogeneration in the year 2000, based on the assumption that electricity is produced 90% of the time that steam is produced.[29]

If we assume a capacity factor of 55%, which is conservative compared with a capacity factor of 65%, used in the Williams report, then:

$$208 \text{ GW} = (208 \times 10^6 \text{ kW}) (.55) = 114 \times 10^6 \text{ kW}$$
$$= 10.0 \times 10^{11} \text{ kWh}$$
$$= 10 \text{ quads}$$

That is to say, 10 quads of electricity can be produced from cogeneration by the year 2000.

The DOE study came up with slightly smaller figures. It illustrates that a 100% adoption of cogeneration would yield a maximum potential ranging from 2.3 to 8.7 quads. The potential would range from 2.3 to 6.15 quads, given existing patterns of government intervention (figure 2).

The upper end of the range corresponds to (1) steam demand satisfied with a high-powered technical mix of 75% steam turbines, 15% gas turbines, and

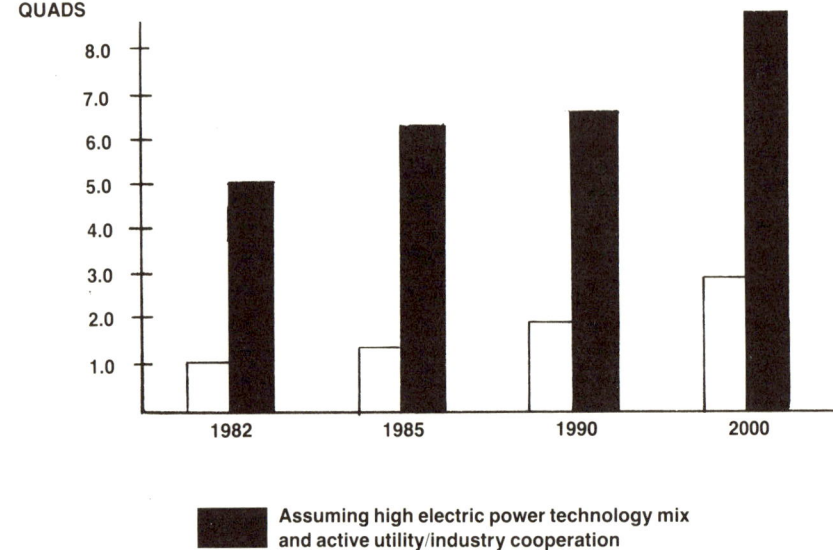

FIGURE 2. Maximum Energy Savings (quadrillion Btu). Source: A. J. Streb et al., *Cogeneration: An Assessment of Commercial Readiness*, Office of Industrial Applications and Commercialization, Department of Energy, 1975.

10% diesel and (2) process heat demand satisfied with 100% gas turbines. The upper end of the range also includes the effect of allowing surplus power to be exploited to the grid.[30]

RPA gave the most conservative figures of the three studies. It analyzed the six major industries in the industrial manufacturing sector: food, textiles, pulp and paper, chemical, petroleum refining, and steel. Eighty to 85% of the total energy used in this sector (approximately 15 quads) is consumed by these six industries. The RPA report estimates a maximum technical potential by 1985 of 3,500 trillion Btu of process steam, which is equivalent to approximately 3.5 quads (table 8).[31] A 2% increase per year in energy consumption in the industrial

TABLE 8. Total Industrial Energy Perspective of Cogeneration Potential

	Industrial Fuel Consumption in 1976 (trillion Btu)					
	Total Fuels Consumed			Fuels Used for Steam Generation	Fuels Used for Process Heat	
Industry Segments	Purchased Fuels	Process Residuals	Purchased Electricity[a]			Other
Six industries	7,900	2,700	1,000	5,400	4,200	2,000
Other industries	14,200	—	1,500	7,700	4,800	3,200
Total	22,100[b]	2,700	2,500	13,100[b]	9,000[b]	5,200

THE POTENTIAL FOR DIVERSITY

TABLE 8 (Cont'd)

	Industrial Process Energy Consumption (trillion Btu)					
	Process Steam		Direct Process Heat		Process Heat Rejection	
Industry Segments	1976	1985	1976	1985	1976[b]	1985
Six industries	**4,100**	**5,900**	1,300	1,900	8,600[c]	10,300
Other industries	5,600	8,000	1,400	2,100	12,500[c]	15,200
Total	9,700	13,900	2,700	4,000	21,100[c]	25,500

	Process Energy Technically Suitable for Additional Cogeneration[d] (trillion Btu)					
	Process Steam		Direct Process Heat		Process Heat Rejection	
Industry Segments	1976	1985	1976	1985	1976	1985
Six industries	**2,200**	**3,500**	750	1,050	2,300	2,760
Other industries	280	400	460	575	650	780
Total	2,480	3,900	1,210	1,625	2,950	3,540

	Technical Limit to Additional Cogeneration Development in 1985			
	Process Steam	Direct Process Heat	Process Heat Rejection	Total
Electric energy (billion kWh)	**168–384**	173–303	35.4–51.4	**380–740**
Capacity (thousand MW)	**24.0–54.8**	24.7–43.2	5.0–7.3	**50–100**
Energy savings (trillon Btu)	**460–1770**	830–1200	370–530	**2200–3500**
(thousand BPDE)	**450–835**	390–565	175–250	**1020–1650**

Note: All numbers, except those in boldface, have been very roughly estimated in an attempt to put the results of this study in perspective. These estimates should not be considered authoritative without further analysis. Numbers are rounded and assume an 80% annual capacity utilization.
a. Estimate based on 1976 data from Edison Electric Institute.
b. Estimates based on the following sources:
"Energy Consumption Fuel Utilization and Conservation in Industry" prepared by Dow Chemical Company for the Environmental Protection Agency (1975); "Industrial Application Study" prepared by Drexel University for the Energy Research and Development Administration (1976); "Efficient Use of Energy," American Institute of Physics Conference Proceedings, No. 25 (1975); "Annual Reviews, Inc." (Vol. 1: 1976); and "A Study of In-Plant Electric Power Generation in the Chemical, Petroleum Refining and Pulp and Paper Industries," prepared by Thermo Electron Corporation for the Federal Energy Administration (1976).
c. Includes heat rejected in producing process steam and heat, heat rejected in the process, and heat lost in other plant operations.
d. In addition to existing cogeneration.
Source: Resource Planning Associates, Inc., for U. S. Department of Energy, *The Potential for Cogeneration Development in Six Major Industries by 1985, Executive Summary*, 1977.

manufacturing sector will yield slightly higher estimates of potential energy savings by the year 2000. Additional savings could also be achieved in sectors not assessed in the RPA study.

Industrial manufacturing sectors such as the stone, clay, and glass sectors, and industrial nonmanufacturing sectors (Standard Industrial Classification [SIC] codes 1–19) such as the agricultural sector, could yield substantial savings from cogeneration systems. We can assume, therefore, that the high technological potential of cogeneration by 2000 will be 6.5 quads—a number within the range of the DOE study yet conservative compared with the Williams estimate and slightly higher than the 1985 RPA estimate. We can also assume a low of 2 quads, which is within the low range of all the assessed studies.

Thus, the total for all cogeneration systems is:

Low   2 quads
High   6.5 quads
Average 4.3 quads = 18% of the 1980 U.S. electricity demand and 159% of the 1980 equivalent of nuclear power generation.

The United States has the infrastructure for developing cogeneration systems. Industries around the Gulf Coast, the Great Lakes, the Middle Atlantic seaboard, and Los Angeles have the best locations for yielding great savings from cogeneration. A savings of 4.3 quads by the year 2000 could alone be enough to reduce the construction of additional centralized power plants.[32] Cogeneration can increase the efficiency of existing systems and help conserve diminishing fossil fuels (figure 3). Another path by which cogeneration could further reduce fossil fuel consumption would involve using alternative fuels such as hydrogen,

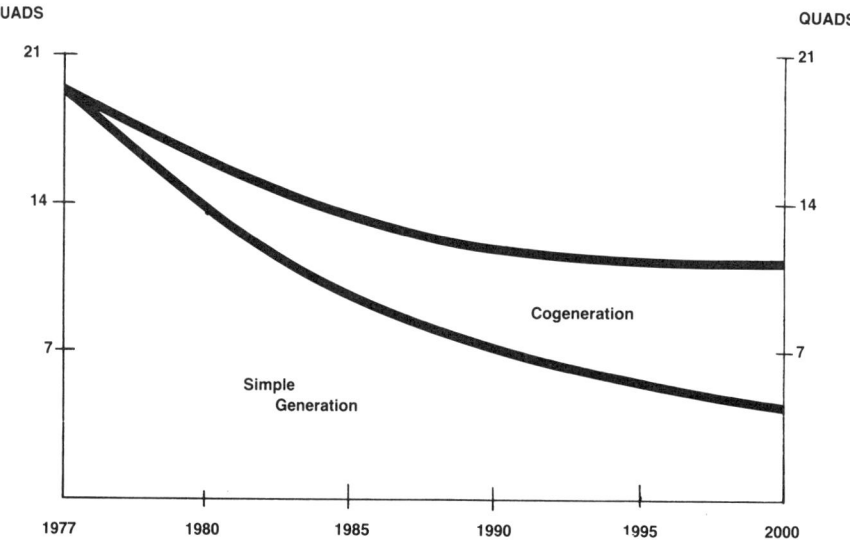

FIGURE 3. Fossil Fuel Use.

THE POTENTIAL FOR DIVERSITY

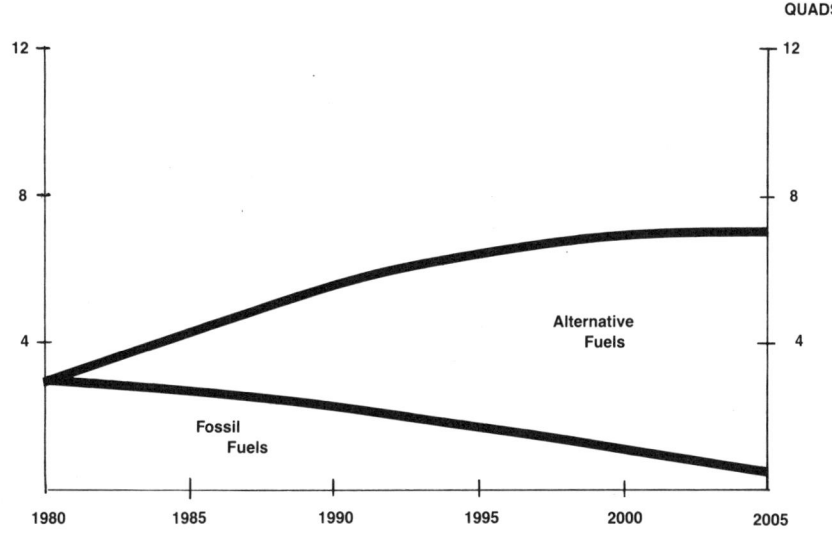

FIGURE 4. Alternative Path for Cogeneration.

methane, and methanol for operation (figure 4). Before the technology can ease the transition to diverse electricity production systems, proper government incentives and regulations must be put into effect.

CUMULATIVE IMPACT

Electricity production from diverse alternative technologies could have a substantial impact on energy supply by the year 2000. Table 9 summarizes my assessment of the cumulative potential impact of the seven alternatives discussed here.

The United States consumed 24.8 quads of electricity in 1980. As table 10 illustrates, there have been various forecasts of electricity consumption for the year 2000. Based on different assumptions, forecast demand figures range from 31 to 96 quads. Recognizing the limitations of forecasting methods, we must set

TABLE 9. Cumulative Impact of Alternative Energy Technologies (quads)

	Low	High	Average
Wind energy conversion	2.2	5.8	4.0
Photovoltaics	1.8	4.3	3.0
Hydroelectric conversion	3.7	4.5	4.1
Solar thermal electric conversion	1.8	3.6	2.7
Ocean thermal energy conversion	1.4	3.5	2.5
Subtotal	10.9	21.7	16.3
Solid waste conversion	.8	1.2	1.0
Cogeneration	2.0	6.5	4.3
Total	13.7	29.4	21.6

TABLE 10. Projections of Energy Consumption (quads)

	Total Energy		Electrical Energy		Average Annual Growth	
					Energy %	Electricity %
	1985	2000	1985	2000	2000	2000
EPP[a]						
Historical	116	187	37	74	3.4	5.5
Tech. Fix	92	124	24	31	1.9	2.2
Zero Growth	88	100	23	31	1.1	2.2
BOM[b]						
Gross	104	163	39	79	3.1	5.5
EEI[c]						
High	116	186	39	86	3.6	5.8
Medium	107	161	36	75	3.0	5.3
Low	101	110	34	51	1.4	3.7
EPRI[d]						
(NGR*)						
High	104	196	41	96	4.2	6.5
Medium	97	159	37	78	3.3	5.6
Low	93	146	36	69	3.0	5.1
(NNGR†)						
High	104	195	40	91	4.2	6.2
Medium	97	158	37	78	3.3	5.3
EIA[e]		(1990)		(1990)		
High	97	109	32	38	—	3.5
Medium	95	109	32	38	—	3.7
Low	92	101	31	36	—	3.0

*NGR = natural gas restrictions; †NNGR = no natural gas restrictions. All figures assume a heat rate of approximately 10,000 Btu/kWh.
a. EPP = Energy Policy Project (Ford Foundation), *A Time to Choose: America's Energy Future*, 1974.
b. BOM = Bureau of Mines, *U.S. Energy through the Year 2000*, 1975.
c. EEI = Edison Electric Institute, *Economic Growth in the Future*, 1976.
d. EPRI = Electric Power Research Institute, *Demand '77*, 1978.
e. EIA = Energy Information Administration, *Annual Report to Congress: 1977*, 1978.

a reasonable goal for electricity consumption by the year 2000. The Council on Environmental Quality's 1979 report, *The Good News about Energy*, claims:

> simple extrapolations of historical energy growth "showed" that the U.S. would need to more than double its current energy consumption by the year 2000. Revised and more realistic estimates now indicate that with a moderate effort to improve energy productivity, our energy consumption in the year 2000 need not exceed current use by more than about 25 percent, and that with a determined effort it need not increase by more than about 10–15 percent.[33]

THE POTENTIAL FOR DIVERSITY

It would be nearly impossible, and ridiculous, for the United States to consume from 50 to 96 quads of electricity. Other countries, such as Sweden and West Germany, have demonstrated that it is possible to maintain a high standard of living without consuming high levels of electricity per capita. By implementing strict conservation measures and strong government incentives, these two countries in the past few years have surpassed the United States in gross national product (GNP) per capita while maintaining a considerably lower energy consumption level per capita. Thus, the United States should set a goal to reduce per capita electricity consumption while converting to alternate energy technologies.[34]

> Some of the main factors that have accounted for the reduced energy use in Sweden are smaller automobiles, more use of mass transit, more insulation and tighter building construction, more efficient industrial processes, and the use of cogeneration and district heating. . . . Higher efficiency accounts for the largest portion of lower energy use in Sweden arising from these factors.[35]

Electricity consumption in the United States would decrease dramatically if there were a shift to less energy-intensive activity and more enforcement of conservation measures. Assuming more efficient use of energy, electricity consumption by the year 2000 should reasonably not exceed 30 quads (roughly 5 quads, or 20%, more than current consumption). A high or average growth scenario for developing and implementing the different electricity production technologies could satisfy a demand for 20 quads of electricity by the year 2000. The studies also indicate that almost 30 quads could be provided by alternative technologies if a strong social commitment were made to the alternatives described here.

Much of the electricity that will be required by the year 2000 can be supplied by diverse alternative energy systems, even if the state of the art of each technology showed no improvement. Existing nuclear and some coal or domestic oil facilities, however, still need to be incorporated with the alternatives to satisfy interim energy demands (figure 5). There are other potentially viable energy-conserving technologies, such as solar space and water heating, wave and tidal power, and small-scale cogeneration (not covered in this chapter), that would further increase the potential of alternatives. I have presented a preliminary assessment of only the most developed technologies. With the added energy contribution of other alternatives not assessed in this chapter and stricter conservation measures, the need for conventional power sources would diminish even further. The unanimity of all the studies in concluding that alternatives are technically available and would have a potentially great impact can only dictate a serious reevaluation of continued investment in fossil and nuclear energy for producing electricity. Decisions of government officials must now favor a more rapid development of alternatives. To eliminate the energy crisis, "there are no simple solutions . . . only intelligent choices."

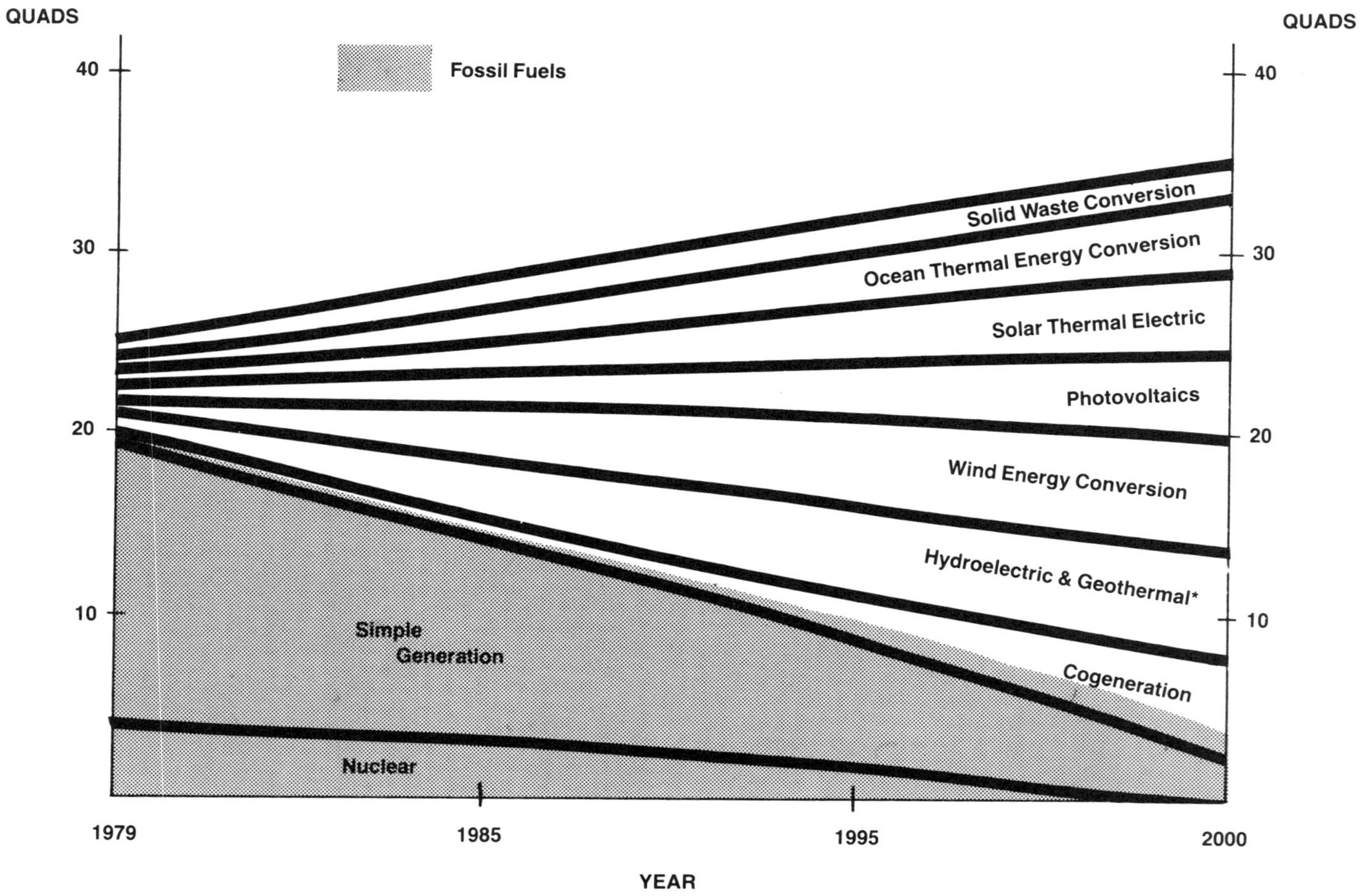

*Assuming no expansion of existing geothermal production

FIGURE 5. A Perspective on U.S. Long-Term Electricity Supply.

NOTES

1. U.S. Department of Energy, Energy Information Administration, *Monthly Energy Review*, March 1981 (Washington, D.C.: GPO), DOE/EIA 0035/81(03).
2. U.S. Department of Energy, *Domestic Policy Review of Solar Energy*, February 1979, TID–22834, table 8.
3. U.S. Department of Energy, *Secretary's Annual Report to Congress*, January 1980, DOE/S–010(80), p. 3–2.
4. Marshal F. Merriam, "Wind Energy Use in the U.S. to the Year 2000," *Wind Energy Report*, August 1978, p. 6.
5. Council on Environmental Quality, *Solar Energy Progress and Promise* (Washington, D.C.: GPO, 1978); Federal Energy Administration, *Project Independence Final Task Force Report—Solar Energy*, 1974, sec. IV; Glenn Reed, "Possible Impact of WECS and Photovoltaics on U.S. Electrical Consumption in 2000" (Paper delivered at the College of Science in Society, Wesleyan University, Middletown, CT, 1978).
6. Arthur D. Little, Inc., *Near-Term High Potential Counties for SWECS*, February 1981 (Golden, CO: Solar Energy Research Institute), SERI/TR–98282–11.
7. CEQ, *Solar Energy Progress and Promise*, p. 9.
8. A very conservative estimate based on an output of .25 kW/m^2 used in John C. Fan, "Solar Cells: Plugging into the Sun," *Technology Review*, August 1978, p. 14.
9. Allen L. Hammond, "Photovoltaics: The Semi-conductor Revolution Comes to Solar," *Science*, July 1977, p. 445.
10. Based on production and deployment assessments from FEA, *Project Independence*, p. VII B–2.
11. U.S. Department of Commerce, Bureau of the Census, *Statistical Abstract of the U.S.* 1977, p. 775.
12. Richard J. McDonald, *Estimate of National Hydroelectric Power Potential at Existing Dams* (U.S. Army Corps of Engineers, Institute for Water Resources, 1977), p. 1.
13. CEQ, *Solar Energy Progress and Promise*, p. 24.
14. DOE, *Monthly Energy Review*, p. 6.
15. David E. Lilienthal, "Lost Megawatts Flow over Nation's Myriad Spillways," *Smithsonian*, September 1977, p. 83.
16. Helen S. Kassler, "Power from the Streams," *Solar Age*, July 1978, p. 17.
17. Daniel Deudney, "Rivers of Energy: The Hydropower Potential," *Worldwatch Paper* 44, June 1981, p. 35.
18. FEA, *Project Independence*, p. III–16.
19. Ibid., p. III–A–12.
20. CEQ, *Solar Energy Progress and Promise*, p. 3.
21. FEA, *Project Independence*, p. III–A–11.
22. CEQ, *Solar Energy Progress and Promise*, p. 26.
23. FEA, *Project Independence*, p. VI–1.
24. CEQ, *Solar Energy Progress and Promise*, p. 26.
25. U.S. Environmental Protection Agency, *Fourth Report to Congress: Resource Recovery and Waste Reduction* (Washington, D.C.: GPO, 1977), 720–116/57, p. 19.
26. Ibid., p. 7.
27. Amory B. Lovins, "Energy Strategy: The Road not Taken?" *Foreign Affairs*, October 1976, p. 65.
28. Robert H. Williams, *Industrial Cogeneration* (Princeton: Center for Environmental Studies, 1978), p. 75.
29. Ibid., p. 71.
30. A. J. Streb et al. for Office of Industrial Applications and Commercialization, U.S. Department of Energy, *Cogeneration: An Assessment of Commercial Readiness*, 1975, part II, p. 4.

31. Resource Planning Associates, Inc., for U.S. Department of Energy, *The Potential for Cogeneration Development in Six Major Industries by 1985*, Executive Summary, December 1977, RA–77–1018.

32. U.S. Department of Energy, *Cogeneration: Technical Concepts, Trends, Prospects* (Washington, D.C.: GPO, 1978), DOE–FFU–1703.

33. Council on Environmental Quality, *The Good News about Energy* (Washington, D.C.: GPO, 1979), p. V.

34. U.S. Department of Commerce, Bureau of the Census, *Illustrative Projections of State Populations, by Age, Race, and Sex: 1975–2000*, March 1979, p. 8.

35. Lee Schipper and Allan Lichtenberg, "Efficient Energy Use and Well Being: The Swedish Example," *Science*, vol. 194, December 1976, p. 1012.

ROBERT D. MORRIS

CHAPTER 5

The Electrical Energy Production System in Transition: The Critical Factor of Reliability

The evolution toward a stable, redundant, renewable electrical energy grid raises legal, political, economic, ideological, environmental, and technical issues. In this chapter Morris introduces some of the technical problems related to electricity production and assesses the possibilities for maintaining reliability in a grid powered by variable sources. After exploring batteries and the possibility of wide-scale independence from the grid, he returns to the central grid concept and explains how management of the grid itself can ensure reliability. Morris gives a detailed discussion of the technology of integrating variable energy conversion systems and also treats the economics of load management and the need for planning in a decentralized electricity system.

The fossil and nuclear fuels that account for almost all our current energy consumption are limited by their very nature. By the simple mathematics of subtraction, the continued use of these fuels will lead to their eventual depletion. Conservation measures can alter the rate of consumption but not the inevitability of depletion. Until we can restructure our energy conversion systems to operate by using renewable energy resources, we will be confronted with continual "energy shortages." A shortage is perhaps a misnomer for the true status of our energy supplies, as it implies a temporary state of affairs with an eventual return to normality. Energy policies that do not include some kind of restructuring or transition will alter only the timing of these shortages; the result will be the same.

The transition to renewable energy resources will require the concerted efforts of planners and policymakers. A renewable energy conversion and delivery system will not be a direct substitution for the existing system. Rather, the transition will be a process of continual integration of new systems into the ex-

The author gratefully acknowledges the aid provided by Dr. Frank Nardine of the University of Wisconsin, Milwaukee, in the preparation of this manuscript.

isting one, with the balance between renewable and exhaustible resources slowly shifting until renewable resources dominate the balance. In this chapter we will focus on an analysis of the issues involved in the shift as it relates to electrical energy generation. The shift process connected with the generation of electrical energy is applicable, with some modification, to power generated from other energy sources of a nonrenewable nature.

THE DEVELOPMENT OF THE UTILITY GRID

That electricity is an awesome and versatile form of energy is often overlooked by the average citizen. The mere flick of a switch sends electric currents racing to obey human commands. Besides being easy to transport, electricity is readily convertible into almost any other form of energy. An appropriately connected wire instantaneously provides access to any device that spins, glows, or flashes. This flexibility has resulted in a steady growth in the demand for electricity, which has brought with it a steady increase in the size of generating plants. Larger plants are the result of a quest for both economies of scale and greater thermodynamic efficiency. The mobility of electricity allows this increase in size by permitting the power produced to be distributed through larger and larger networks.

As the plants have increased in size, so have the distribution networks associated with them. The networks, or combined networks known as grids, have begun to overlap, thereby interconnecting several power plants and stretching over vast areas. The process of expansion and interconnection has continued at a steady pace, and today power is sent hundreds of miles at several hundred thousand volts. Indeed, national and even international utility grids are seen by many as inevitable.

THE ILLUSION OF ELECTRICITY'S RELIABILITY

To the average consumer the intricacy and sophistication of power plants, transmission lines, and interconnecting grids are generally taken for granted. The exact source of the energy from the wall outlet may not be given more than a passing thought. Further, the seeming simplicity of a light socket betrays the complexity of the various components and their interrelationship which makes electricity readily available. Inherent in the illusion of simplicity is an associated illusion of reliability. That is, we take it for granted that electricity is "in the wall" when we plug in. What is the consumer's predictable reaction should he or she suddenly find no energy in the outlets throughout the house? Surprise? Indignation? Consternation? Inevitably a call to the utility. Although far too many citizens cannot contemplate the prolonged unavailability of electricity, in part because their life-styles would be drastically affected, each of us will become increasingly aware that there are limits to the energy sources currently supplying our electricity. Thus, the reliability of electricity is very much contingent upon the particular source of energy that generates it.

Oil, coal, natural gas, and nuclear fuels account for almost all our current electrical supplies. The limited quantity of oil and natural gas reserves is becoming well known. We must also realize that both have far more value as fuels for transportation and heating needs than as fuels for the production of electricity. At present coal is far more plentiful than oil or natural gas, particularly in the United States. However, the environmental drawbacks to both the mining and burning of coal would be greatly magnified by a substantial increase in the rate at which we consume coal. Furthermore, coal resources are also finite. An increased dependence on coal will offer only temporary relief from our energy problems. For a while nuclear power seemed promising, but questions about its safety and economic viability make a commitment to it an unsettling alternative. Like fossil fuels, nuclear fuels are a limited resource and cannot be infinitely reliable.

Reliability and Energy Source
The plentifulness of an energy source and the degree of reliability we place in power generated from it are related. Exhaustible resources, which supply most of our electricity, have short-term reliability. The energy available in these resources is stored. Fossil fuels hold the energy of millions of years of sunlight, while nuclear fuels hold the primordial energy of unstable matter. Because this energy is stored, it can be released whenever it is desired. The only apparent limitations in the short term are the rate of production and the availability of conversion facilities, namely, power plants. However, once this energy is released, we have no way to replace these fuels. The only energy source with assured long-term reliability is the sun. The sun annually bathes the earth with energy equivalent to our entire remaining supply of fossil and nuclear fuels. It creates a flux of energy that is manifested in many forms: wind energy, biomass energy, ocean thermal energy, and hydropower, as well as direct radiant energy, are all forms of solar energy.

Hydropower, ocean thermal power, and biomass energy provide various forms of stored solar energy and, as such, can provide a stable and consistent flow of energy. By far the most plentiful and widely available forms of solar energy are wind and direct radiant energy. But the incident wind or solar energy at any one location is highly variable over short periods of time. Although these are extremely promising sources of energy, they therefore lack the short-term reliability of supply on demand that we now enjoy from other sources.

Historically, the question of long-term reliability has not been discussed to any significant extent because of the nature and seemingly abundant supply of fossil fuels, which have generated our electricity. Only recently has the cost of this omission become painfully clear. Should we, in the not-too-distant future, be able to shift to a dependence on solar energy, we may be successful in addressing the question of long-term reliability, but this shift would almost instantly force us to confront the problems of short-term reliability. More specifically, the ceaseless flow of solar energy to the earth varies in form and intensity with respect to time, as we may observe at any one location. The variability and resulting lack of reliability of whatever sources of energy we utilize must become, therefore, a forethought rather than an afterthought in the process of mapping out energy production and consumption.

THE USE OF ENERGY STORAGE TO ACHIEVE SHORT-TERM RELIABILITY

The inexhaustible nature of solar energy has spawned visions of energy-autonomous homes, businesses, and communities, free from power lines and endless monthly billings. In reality, this alternative is not so simple. Suppose we would like to power a small dairy farm with a photovoltaic array. An array can be designed such that its average output exactly matches the average demand of the farm, but the production of the array at any moment will rarely match the electrical demand of the farm. In order for our photovoltaic power system to approximate the reliability of fuel-based power systems, we must somehow redistribute the energy output of the photovoltaic power system so that energy produced in excess of demand can be made available during subsequent periods of production shortfall. This temporal redistribution of energy can be accomplished through electrical energy storage.

Electrical energy storage systems allow energy produced in a limited time frame to be stored, thereby extending the time frame during which it is available. Electrical energy storage has come to represent the ideal and simple solution to the problems posed by variable energy sources. In general, the energy dilemma does not yield to simple solutions, and this pattern is not disrupted by energy storage. At present, no universally applicable electrical energy storage technology is economically viable. Major research efforts for both electric vehicles (Behrin and Anderson 1978) and utility peak load shaving (Birk and Pepper 1977) have been undertaken, but significant breakthroughs have not yet occurred.

Two energy storage technologies with implications for electricity have already reached economic viability: pumped water storage and compressed air storage. However, siting requirements limit the applicability of both technologies. Briefly, pumped water storage involves pumping water into a reservoir to be used later for the production of hydroelectric power. In a compressed air power system, energy can be stored by compressing air and storing it for later expansion through turbine generators. Both technologies require a suitable location; a pumped water storage system must have an appropriate site for a storage reservoir, while a compressed air system requires an underground cavern or some other geological formation capable of containing large volumes of compressed air.

A number of other storage technologies under consideration do not have siting constraints. The three most promising are advanced battery systems, hydrogen generation and conversion systems, and inertial storage systems. Of the three, advanced battery systems show the greatest potential for achieving commercial viability in the near future. The common lead-acid battery is currently too expensive to be cost-effective, but research on a wide range of new battery types (Robinson 1976) shows some promise. In ten to twenty years advances in battery design and function may provide an economically viable electrical energy storage option. Hydrogen-based systems and inertial storage systems (flywheels) do not have major near-term potential but they may be valuable options in the long term. Experimental flywheels (Biggs 1974) are designed to spin at extremely high rotational velocities in evacuated chambers, but these "super-flywheels" have yet to be fully proven as a technology. Initial tests indicate that they have potential (Hagen, Erdman, and Frohib 1979), but far more work needs to be done. The electrolytic generation of hydrogen and oxygen from water shows unique potential in that the end product, hydrogen, is a fuel and, as such, has a broad range of possible applications besides the production of electricity. Gas stoves, water heaters, and tractors can all be altered to run directly on hydrogen. At present, the cost of equipment for the generation and storage of hydrogen and the subsequent regeneration of electricity from hydrogen is extremely high, but numerous breakthroughs, such as the use of lower-cost photovoltaics, could change this. The primary goal of researchers is to reduce these costs (Braun et al. 1976).

Electrical energy storage appears now to be a prohibitively expensive and impractical option for our solar-powered farm (Applied Research 1978). Even pumped water and compressed air storage systems are economical only as large utility storage facilities.

THE USE OF THE UTILITY GRID TO ENSURE RELIABILITY

How, then, can we provide reliable power to our solar farm? Perhaps we should not be so hasty in divorcing ourselves from the existing utility network. With the exception of a relatively small number of remote sites that must use expensive lead-acid batteries or gasoline generator sets, nearly all the electrical loads in the United States have access to a utility grid. Utility power lines make the stored energy of fossil fuels available to these loads. By interconnecting our hypothetical solar-powered farm to the utility grid, the farm's power supply will be as reliable as the utility power system. An electrical energy storage system is no longer necessary as a method for eliminating variability. In this system, the photovoltaic array serves to reduce the farm's consumption of power from the utility whenever the sun is shining. The array is, in essence, an energy-conservation system.

In the absence of energy storage, however, we still have no way to absorb excess production. If some portion of the array's output is unusable, the total effective output of the system decreases. As a result, the energy cost for the system increases.

The unique mobility of electricity allows the almost instantaneous geographical redistribution of excess electrical production through the utility grid. This electricity can then be absorbed by other loads on the same grid. Thus, the existing infrastructure for the production and distribution of electricity both provides a method for eliminating variability in the flow of power to our solar-powered farm and allows for complete utilization of the output of the variable energy conversion systems (hereafter referred to as VECS).

The concepts underlying the interconnection of VECS with the utility grid can be explained by substituting the flow of water for electrical currents. In this way we can develop an analogy for grid integration in accessible, macroscopic concepts. The utility grid, as it operates today, can be compared to a water distribution network consisting of several large pumps feeding water into a series of continuously branching conduits. Let us consider our decentralized VECS to be a pump operating at the end of one of these branches. Like a VECS, this pump would have a variable output. To complete the analogy, let us place a load such as an irrigation system next to our small variable pump. When the small pump is inactive, water will flow into the branch in accordance with the water requirements of the irrigation system. As the small pump begins to function it will supply whatever portion of the load it can, and as a result the amount of water running from the central pump will decrease. As the output of the pump increases, it will first meet and finally exceed the water requirements of the irrigation system. At this point, flow into the branch will have stopped and excess water will begin to flow out of the branch. This water will join the flow of water in the larger branch in supplying other, smaller branches. As more small pumps are added to nearby branches, their output will, at times, be sufficient to push water into larger and larger branches.

TECHNOLOGIES FOR INTEGRATING VECS AND THE UTILITY GRID

Four basic technologies can be used to integrate VECS with the utility grid: (1) synchronous generators; (2) induction generators; (3) line-commutated inverters; and (4) self-commutated inverters. Economic and engineering considerations determine which technology is optimal for a specific application. In a properly designed system the output of the VECS matches the utility power in voltage, frequency, and phase. If it does not, the interconnection causes disruptions in the grid and the power produced is partially or completely unusable.

SYNCHRONOUS GENERATORS. Synchronous generators, like induction generators, require a source of rotating mechanical energy to operate. This energy can be provided by wind turbines, hydro turbines, or steam turbines. Essentially, all conventional power plants use synchronous generators to convert the mechanical output of a steam or hydro turbine into electricity. A synchronous generator produces AC power at a frequency directly proportional to the rate at which the generator turns. By controlling the heat rate and/or flow of steam or water to the turbine, the speed of the generator can be held constant under varying load conditions and the output frequency can be maintained at a constant speed.

Controlling the rotational speed of a generator by adjusting the flow rate through a turbine is not feasible when the working fluid is the wind. A wind turbine or, for that matter, a small hydro or solar thermal turbine, which is susceptible to momentary variations in the energy influx, may be incompatible with synchronous generators.

INDUCTION GENERATORS. An alternative to synchronous generators for direct connection to the AC line, an induction generator is nothing more than an induction motor that is driven at speeds slightly greater than synchronous speed (60 Hz output). Although the rotational speed must be held within limits, there are far fewer constraints than for a synchronous generator. Speeds as much as 5 percent greater than synchronous speed are acceptable.

LINE-COMMUTATED INVERTERS. The use of a solid state inverter eliminates the need for precise rotational speed controls insofar as interconnection is concerned. An inverter can produce AC power compatible with the utility power given a variable DC input. Some energy conversion technologies, such as photovoltaic cells or fuel cells, produce only DC power and cannot interconnect to the utility without an inverter.

Line-commutated inverters are variable DC motor drives operated in reverse. Rather than producing variable DC power from the AC line, they produce line-compatible AC power from variable DC power. The inverter is "line-commutated" because it requires a signal from the AC line for power to flow.

Both induction generators and line-commutated inverters are line-dependent technologies in that they cannot produce AC power in the absence of an external AC source. These two technologies are less expensive and simpler to use than systems capable of operating independently, such as synchronous generators and self-commutated inverters.

SELF-COMMUTATED INVERTERS. A self-commutated inverter is a DC-to-AC converter that uses an internal oscillator to build a waveform of the desired frequency. Currently, the most common application for this sort of system is in uninterruptible stand-alone power supplies for computer facilities. The use of these inverters as line-feeding inverters is a relatively new application.

THE ECONOMICS OF INTERCONNECTION

To varying degrees, all these technologies have already been used to feed power into utility grids. Thus, as we begin to integrate variable sources into the grid, the major issues will not be technological but economical.

These technologies are far less expensive than energy storage systems. By serving the same function as storage—the minimization of variability—grid integration can improve the economic viability of VECS. The power from the VECS will first flow to the load dedicated specifically to the VECS while excess power is fed into the grid and distributed along utility power lines. The value to the owner of the energy produced will depend on which of these two courses the energy follows and, if power is fed back to the utility, what policy of payback the utility maintains for purchasing this power.

Utility payback policies are now and will continue to be developed over the next few years (see chap. 7), and the nature of these policies will have a critical effect on the economics of dispersed variable sources. Section 210 of the Public Utility Regulatory Policies Act of 1978 requires utilities to formulate equitable payback policies. The utility payback rate for electricity fed into the utility grid will reflect only the cost of generating replacement power, and not transmission and distributing costs. This power would then be sold to the utility at wholesale cost and purchased from the utility at retail cost. As a result, the most economically attractive situation for the application of a variable power source is one for which the dedicated load always or almost always exceeds the maximum power output of the variable source. All the power produced by such a system would thus have the full retail value of utility power.

VECS AND UTILITY PLANNING

In planning power generation, the utility is concerned only with the total load placed upon the system. Power flowing into the line from an interconnected VECS is absorbed by other loads, thus serving to reduce the total load to be met by central plants planned by the utility. The output of a VECS varies just as the power consumption of any utility load might vary. Thus, the effect of VECS on utility planning is that of a net negative load.

The first few VECS installed in any grid system will have a minimal effect on the total load and will probably not receive special attention in utility planning procedures. However, as penetration (that is, the generating capacity of VECS as a percentage of the total capacity of the utility) increases, utilities will be forced to involve these systems in their planning process. A study modeling the introduction of wind systems into a Swedish utility indicated that a 10 percent penetration level could be sustained without including the wind in generation planning (Larsson 1978). The penetration level at which planning becomes necessary will vary depending on the characteristics of the individual utility. Initially this planning may amount to little more than evaluating the wind/solar resource on a daily basis for the following day and allotting conventional generating capacity accordingly. In this scenario the utility would need to maintain sufficient conventional generating capacity to meet any system load. The output of the VECS would serve only to reduce fuel consumption or water usage at thermal or hydroelectric generating plants. The interconnected VECS would not reduce utilities' generating capacity requirements. If we are to replace conventional generating plants with VECS, we must improve the reliability of VECS to the extent that some portion of their total generating capacity can be counted upon to supply electricity as needed.

THE RELIABILITY OF AN INTERCONNECTED NETWORK OF A DISPERSED VECS

To improve the reliability of each individual system integrated with a utility would require energy storage. The expense of this option dictates that we exhaust other available options first. This requires considering the reliability of the overall grid system rather than just one generating point on it. Recall that wind and solar energy vary not just with respect to time for a given location but with respect to location for a given time. By increasing the area from which we draw energy, we increase the likelihood that at any given time, somewhere in the utility grid, power from variable sources is available to us. In this way we effectively improve the reliability of each VECS without additional cost. Studies (Molly 1977; Justus and Mikhail 1978) indicate that as this grid increases in size, the level of correlation and therefore the effective reliability of the total VECS input increase with the number of sites and the area over which they are spread.

Lulls in the available wind power tend to occur during the summer in temperate climates, owing to reduced atmospheric temperature differentials. Radiant solar energy, on the other hand, is at its peak during the summer and at its lowest during the winter, when winds are highest. The complementary nature of wind and solar energy will further improve the utility-wide reliability of VECS systems. Computer models (Andrews 1976) tend to support this contention. In the future, utilities may extend their transmission lines with the specific intention of including variable sources with low levels of correlation.

Both Molly and Justus, in studying the reliability of dispersed wind systems, examine reliability and expected power levels without regard to when those power levels occur. Their analysis does not allow for the fact that the utility load is variable. In fact, the average daily peaks for both wind and solar energy match closely the average daily peak loads for the utility. All three occur during the middle of the day. Thus, the lulls that do occur in the combined wind/solar output will tend to occur during periods when demand is low. This indicates that at least some variable sources may be useful in displacing valuable peak energy supply.

LOAD MANAGEMENT AS A TOOL FOR INCREASING RELIABILITY

The ultimate test of reliability is the ability of a power network to meet demand. Although the dispersion of VECS over a broad area will reduce the extent and frequency of variation, it is inevitable to some degree in a grid with moderate to high VECS penetration. The high correlation between power output and demand may reduce the impact of this variation on reliability, as we have defined it here, but to some degree noncoincidental supply and demand will be inevitable given current consumption patterns.

Traditionally, power output has been adjusted to match largely uncontrolled variations in demand. By managing consumption patterns, it may be possible to improve dramatically the reliability of our VECS network. In the United States most attempts to adjust demand have been aimed at increasing it by charging lower average rates to larger consumers. Declining block rates, as this type of rate structure is called, have historically been the most common residential electrical tariffs in America. In other industrial economies energy costs have usually been higher than in the United States. The result of this, as three economists associated with the Rand Corporation point out, is that "foreign utilities have taken a more active approach to customer loads. They have not accepted the existing load pattern as a given, but have tried to reduce costs by actually shaping the system load" (Mitchell, Manning, and Acton 1978). Rising energy costs in the United States have drawn increasing attention to the potential for rate reform in this country (Cicchetti, Gillen, and Smolensky 1977). As these costs continue to rise, foreign load-management strategies will play an expanding role in American utility planning.

The current role of load management in utility systems is primarily the reduction of peak loads. Some utilities with high hydroelectric penetration have sought to reduce average demand during periods of low water, but for the most part utilities seek to decrease peak demand as a way to lower the average costs of generating electricity. These cost reductions result from decreases both in the total generating capacity requirements of the utility and in the need to use existing peaking plants which are usually fired by oil or gas. In the case of high VECS penetration, the goal of load management would be to minimize the differential between VECS power output and the coincident utility load. This differential will define the need for supplemental power systems, either electrical energy storage or fuel-fired generating plants. By reducing the requirements for additional equipment, as well as fuel, we can reduce the total cost of the system and thus the cost of the energy produced. The utility load must be altered as far as possible to match system-wide variations in VECS power output. The variations in the output of an integrated network of wind and direct solar energy conversion systems can be classified as either seasonal variations in average energy production, continuous low-amplitude oscillations in power output, or occasional extremes in the system power output. Each of these characteristics suggests a different load-management strategy.

Seasonal Load Management
The nature of seasonal variations in energy production will depend upon the mix of wind and solar sources in the utility grid. The system may experience lows in average power output either in summer and fall (higher wind penetration) or in fall and winter (higher solar penetration). If the system is balanced between wind and direct solar energy, fall will tend to be a period of low average output. Various utilities have used seasonal pricing structures either to discourage the use of seasonally variable loads, such as air conditioners, or to account for low hydroelectric availability during dry seasons. With little or no alterations, seasonal rates could be used to discourage electrical energy consumption during periods of low availability and encourage use during periods of high availability of wind and solar energy. The exact period will depend on the utility's total generating mix and the annual weather patterns for the area. When twenty-five VECS locations in the central United States were integrated into a hypothetical network of wind systems (Justus 1978), the network produced at least 43 percent of its average output 90 percent of the time during the month of January but reached the same level only 77 percent of the time in October. High rates during October would reduce both average and peak consumption for this period. Thus, the capacity required to meet the load during periods

of low wind availability could be reduced, as could the consumption of fossil fuels.

Buffers to Absorb Minor Variability
Within the annual patterns of variation in wind and solar availability are continuous low-amplitude momentary and daily fluctuations in power output. We can minimize the effect of these short-lived oscillations by placing a buffer between the power input from the utility and certain readily adaptable elements of the system load. This buffer would be capable of absorbing momentary excesses in the energy production and transferring the energy to subsequent periods of low production. A prime example of this sort of buffered load is an electric water heater. Most electric water heaters have storage tanks and can absorb substantial amounts of energy. The water can then be stored for several hours. The power flow to these heaters can be switched on and off many times in the course of a day without disrupting the availability of hot water. If this switching operation is controlled by the utility, the utility will have some capability to regulate the system load to match the available power. Water heaters with remote switching capabilities are available commercially and have been tested by a number of utilities in the United States (Schaefer 1979). A recent study (De Winkel 1979) evaluated the potential for replacing the peak-load generating capacity of a Wisconsin utility with wind-generated electricity and storage water heaters. The peak-load generating capacity supplies the top 15 percent of the utility's peak load. A wind-power capacity equal to the conventional peak-load capacity and a hot-water storage capacity equal to 23 percent of the peak-load capacity met the load about 92 percent of the time over a five-year period, "a reliability similar to or better than a conventional peak load generator."

Other loads can be buffered in a similar manner. England and Wales have installed 15,000 MW of clocked controlled storage space-heating equipment over the past twenty years (Mitchell, Manning, and Acton 1978). Cold storage is also under investigation as a load-management strategy ("Cold Energy Storage" 1979).

Compressors designed for use with pneumatic tools store some volume of compressed air in a pressure vessel. By increasing the size of the storage tank, a buffered load could be designed for use with power tools.

Strategies and Incentives for Changing Daily Load Patterns
Buffered loads can be used to alter consumption patterns as seen by the utility without shifting use patterns as viewed by the consumer. By providing the customer with appropriate incentives to shift consumption patterns of nonbuffered loads, utility demand patterns can be altered still further. The reaction of European industries to rate structures designed to reduce peak loads clearly indicates the ability of consumers to respond to load-management incentives. Tariffs during peak demand periods have induced industries to adjust scheduling for full-scale production and maintenance activities, increase levels of productive off-

peak periods, and generate their own power from industrial by-products. The exact combination of methods employed depends on the processes involved.

The incentives used to effect these alterations in load are not directly applicable for use with our high VECS penetration grid. Because peak loads consistently occur during the middle of the day, peak-load pricing can be based on time-of-day rates. Although, on the average, solar and wind peak power output is during the middle of the day, the actual output will vary from one day to the next. The degree of variation will depend on the size and location of the grid. As a result of this variation, a rate structure based on power availability would be difficult to administer. A system involving a remotely programmable metering system to which industries could respond is conceivable and, although it appears unmanageable, should not be ruled out as a possibility.

A more likely incentive would be the result of industry ownership of VECS. It would be desirable to the industry concerned to utilize as much of the VECS power as possible. As a result, the industry might be enticed to alter consumption patterns without utility involvement.

Load Shedding as an Extreme Measure

Extreme fluctuations in combined VECS output will occur only occasionally, but they will tend to be more problematic than seasonal or regular low-amplitude fluctuations. Extreme reductions in availability are those most likely to cause significant power shortages. In the case of such a fluctuation, drastic load-management techniques would be required to minimize or eliminate the potential power shortage. Interruptible rates and load shedding could prove effective in allowing for extreme reductions in power availability.

Interruptible rates are reduced rates offered to industries in return for allowing the utility to reduce the power available to them during extreme peak loads. These interruptions may occur only a few times during the year, but they reduce the total generating capacity needed to meet peak loads. During periods of extreme shortfall in power availability in a grid with high VECS penetration, the load shedding allowed by interruptible rates can reduce the requirements for supplemental generating capacity. A load-shedding cooperative (*Energy Users Report* 1980) recently organized among four large utility customers in Southern California was able to reduce its peak demand by 25 percent using electronic data processing equipment to monitor and control energy consumption in ten different buildings owned by these customers. By shedding load six times during the first year of operation, the cooperative was able to reduce utility capacity requirements by 4 MW.

Power production shortfalls are by nature more problematic than production excesses, but power availability in excess of total utility demand must be considered and confronted. Excess power can be sold to neighboring utilities through transmission links where practical. In Norway excess hydroelectric power created by summer runoff is sold during off-peak periods at extremely low rates (Mitchell, Manning, and Acton 1978). Similar rates might be made available in the United States during periods of excess wind or solar availability.

If these methods are insufficient, wind and solar sources can be disconnected, but this option should be considered as a last resort.

THE NEED FOR SUPPLEMENTAL POWER SOURCES

It is unlikely that these strategies will completely eliminate the need for a supplemental source of power unless fundamental changes in consumption patterns can be induced. Assuming this does not occur, some source of stored energy will be required during periods of extreme production shortfall. Because of the lower costs involved and because they do not require significant amounts of additional resources, load-management strategies should be exhausted before fuel-fired generating plants or energy storage systems are called for. The costs of these nonvariable power sources may, in fact, determine the degree to which they are employed. Extremely high costs will make further load management more attractive.

While fossil and nuclear fuels can provide supplemental power for the early stages of the scenario we have outlined, electrical energy storage, biomass fuels, and hydroelectric power, along with tidal power, wave power, and ocean thermal power in some combination would fill this function in a fully renewable electrical energy system.

Biomass fuels can be used only to provide supplemental power, but energy storage and, in some instances, hydroelectric systems are capable of redistributing energy over time. In this way, reliability can be increased without a comparable increase in installed generating capacity.

The future viability of the various energy storage technologies is impossible to ascertain at present. By integrating VECS into a grid network, however, we provide access to site-dependent energy storage options that are not available for most on-site energy storage applications. These technologies—pumped water storage and compressed air storage—are economically feasible today but can operate only as centralized utility storage systems.

Innovations in energy storage, whether batteries, flywheels, hydrogen, or some other technology, may further ease the task of maximizing the reliability of VECS. It should, however, be noted that even in the absence of new developments in energy storage technologies, the reliability of our renewable grid can be improved significantly using pumped hydro and compressed air storage in conjunction with load management.

Additional supplemental power can be supplied by biomass fuels (for example, wood, forestry wastes, agricultural wastes, and alcohol) or some emerging technology, such as wave energy systems or ocean thermal energy conversion systems. The exact combination of these technologies to be employed will depend on the need for supplemental power, the location of the utility grid, the future development of these technologies, and, in the case of biomass fuels, the demand placed on these energy sources by other end uses.

Given present energy realities, it is evident that we have lived through the golden age of petroleum and are now entering a period of transition—from almost total dependence upon fossil fuels to a new dependence, we hope, upon renewable forms of energy. The industries capable of supplying the equipment needed to utilize this energy, however, are in their infancy. It is difficult for many people even to contemplate a world in which fossil fuels are utilized solely as emergency rations, to be consumed judiciously in times of drastic need. In effect, fossil fuels and the like would serve as backup for renewable energy sources.

We have examined ways of addressing the issue of reliability in the process of shifting from nonrenewable to renewable forms of energy. We have looked briefly at improving the reliability of the source and at improving overall reliability by adjusting demand to approximate power availability using accepted load-management techniques. Still another aspect of a sustainable energy future remains to be addressed: that of changing energy consumption patterns. It is beyond the scope of this chapter to present a comprehensive examination of this issue, but it is eminently reasonable (though to some unthinkable) to reshape the tools of our society in order to allocate our energy resources with maximum effectiveness. This reshaping process does not necessarily represent a lowered standard of living but rather a somewhat different standard from our present one. Razors and blades can once again be manufactured to last instead of to be thrown away after several uses. Vehicles can be produced to last twenty years instead of five. Learning to reuse can become the rule instead of the exception. Conserving by preserving does not in itself lower our living standard.

The scenario developed above is not a plan for the transition to an electrical energy system based on renewable sources of energy. Rather, we have suggested a framework of possibilities that should be considered in constructing such a plan. It is impossible to develop a long-range plan that anticipates future events with perfect accuracy. Thus, as a plan of this nature is implemented, it must be continually re-evaluated and adjusted in light of new understandings and technological advances.

A long-range energy plan should have sufficient flexibility to adjust to technological breakthroughs, but it must not rely heavily on them for its success. A long-range plan must be based on existing or proven technologies. The tendency to depend upon momentous breakthroughs can lull us into a dangerous complacency, a belief that if we can hold out for a while the ideal alternatives will be found. We cannot afford to wager our future on technological dark horses at the risk of squandering the birthrights of our children. We must begin to entertain the possibility of changing our everyday requirements to preserve an adequate standard of living for future generations.

REFERENCES

Andrews, J. W. "Energy-Storage Requirements Reduced in Coupled Wind-Solar Generating Systems." *Solar Energy*. Vol. 18. Pergamon Press, 1976.

Applied Research on Energy Storage and Conversion for Photovoltaic and Wind Energy Systems. Final Report, Vol. 1. Study Summary and Concept Screening. NTIS U.S. Department of Commerce, January 1978.

Behrin, E. and Anderson, C. J. "Energy Storage Systems for Automobile Propulsion," U.S. Lawrence Livermore Lab Report, UCRL–52553. Vol. 2. Berkeley, December 15, 1978.

Biggs, F. "Flywheel Energy Systems." Sandia Laboratories, SAND 74–0113. November 1974.

Birk, J. R., and J. W. Pepper. "Energy Storage in Electric Utility Systems." A paper presented at Energy Use Management International Conference, Tucson, October 24–28, 1977.

Braun, C., A. Beaufrere, S. Srenevason, G. Strickland, and J. J. Reilly. "Hydrogen for Energy Storage: A Progress Report of Technical Developments and Possible Applications." Brookhaven National Laboratory, Energy Storage Conference, February 1976.

Cicchetti, C., W. Gillen, P. Smolensky. "The Marginal Cost and Pricing of Electricity: An Applied Approach." Cambridge, MA: Ballinger, 1977.

"Cold Energy Storage," *Heating Piping Air Conditioning*. Vol. 51, no. 4, April 1979.

De Winkel, Carel C. "An Assessment of Wind Characteristics and Wind Energy Conversion Systems for Electric Utilities." (Applications for Wisconsin and Sections of Minnesota, Iowa, and Illinois). IES Report 104, DSPE Special Monograph Division of State of Wisconsin, January 1979.

Energy Users Report (A Weekly Review of Energy Policy, Supply, and Technology). Section 1, no. 358. Washington, D.C.: The Bureau of National Affairs, Inc., June 19, 1980.

Hagen, David L., A. G. Erdman, D. A. Frohib. "Design of Flywheels Utilizing Cellulosic Materials." 14th Intersociety Energy Conversion Engineering Conference, August 6, 1979.

Justus, C. G. "Wind Energy Statistics for Large Arrays of Wind Turbines (New England and Central U.S. Regions)." *Solar Energy*. Vol. 20. Pergamon Press, 1978.

Justus, C. G., and A. S. Mikhail. "Energy Statistics for Large Wind Turbine Arrays." *Wind Engineering*. Vol. 2, no. 4, 1978.

Larsson, Lennart. "Large-Scale Introduction of Wind Power Stations in the Swedish Grid: A Simulation Study." *Wind Engineering*. Vol. 2, no. 4, 1978.

Mitchell, B. M., W. G. Manning, Jr., and J. P. Acton. "Peak Load Pricing." Cambridge, MA: Ballinger, 1978.

Molly, J. P. "Balancing Power Supply from Wind Energy Converting Systems." *Wind Engineering*. Vol. 1, no. 1, 1977.

Robinson, A. "Advanced Storage Batteries: Progress but not Electrifying." *Science*. Vol. 192, May 1976.

Schaefer, John C. "Equipment for Load Management: Communications, Metering and Equipment for Using Off-Peak Energy." Topic Paper 4. Electric Utility Rate Design Study, Palo Alto, October 1979.

BENT SØRENSEN

CHAPTER 6 | *The Grid as Energy Absorber and Redistributor*

Sørensen extends the discussion of grid integration to include more details about managing a diverse electricity grid. He explores the problems of load management, base and peak-load facilities, and large and small-scale storage in greater depth. He assesses intermittent sources such as solar electric, wind, and hydropower, showing how each lends itself to base or peak, to storage or grid integration, to a particular time of day or season, and more. This study is wide-ranging, introducing many key issues in utility planning. Sørensen defines a future role for the grid and points out that centralized electricity transmission/distribution systems are appropriate even in a decentralized world. Such grids, he writes, "serve to exchange power in the case of load mismatch, and they may serve as a security in case of failure or other problems of individual [electrical] conversion units." The distinction between supply and distribution is further discussed by Huettner in chapter nine.

In remote regions without access to an extended electrical transmission system, there is an obvious advantage to depending on energy conversion using only local sources. It is natural to ask, however, whether it is still advantageous to use local energy sources when access to a grid system is possible. This chapter surveys some of the advantages and disadvantages of decentralization and of specific conversion systems that utilize local renewable energy sources.

Generally, the presence of a grid allows for locating energy conversion units at optimal sites. In the case of boiler-type generating units (nuclear or fossil), this may apply to the presence of adequate sources of cooling water or to proper safety distances. However, a conflict of interests may arise in defining siting criteria or choosing the optimum size of units, in order, for example, to use reject heat for industrial processes or district heating.

For renewable energy sources, the presence of a grid may allow siting in regions most favored by the available form of energy; for example, wind energy converters may be located in areas with high winds and concentrating solar collectors in areas with little cloud cover. Most renewable energy converters are flexible because of their modular nature, so that the size of a generating plant may start small and the total capacity may be gradually increased as needed. Most renewable systems can also be quite dispersed in comparison with present systems, which tend to operate as large central plants that must be sized once at the outset.

DECENTRALIZED PRODUCTION

At present the function of electrical utility grids is to carry power from a few places of production to a large number of users. An extreme situation of decentralization would be to have about as many producers as users, for example, if the roof of every building were to carry an array of photovoltaic cells.

Partial decentralization involves many possibilities. A situation with one large wind energy converter per church tower and one biogas-fueled generator per farm would constitute a fairly high degree of decentralization, whereas clusters of wind energy or wave energy converters along shorelines or offshore would resemble in some ways the large, centralized units of today while still preserving some aspects of decentralization.

Dispersed power generation generally diminishes the length of transmission. However, because a transmission system is still necessary, no savings on grid construction and maintenance can be expected. On the contrary, some of the decentralized sources of energy envisaged in decentralized systems may require reinforcement of the transmission lines owing to occasional surpluses of production that exceed the peak load demand (which in the absence of decentralized power inputs would be the maximum transmission capacity demanded).

The losses associated with power transmission depend on the distance of transmission and on the type and quality of equipment used, as well as on atmospheric conditions (if overground transmission is used). Typical average transmission losses for utility systems in industrialized regions are currently on the order of 10 percent or less. No more than half this transmission loss could be avoided by a completely decentralized mode of production, and since the entire grid system would still have to be maintained for the provision of back-up power, the potential dollar savings in the area of transmission by switching to decentralization may well be negligible.

The crucial question is whether it is desirable to utilize alternative energy sources, many of which are best used in a decentralized way. The decision can be based primarily on economic or resource considerations, or on the conviction that better societies may be built around a decentralized means of production (of energy or other goods).[1] In order to make the wisest decision, it is essential to gain a better understanding of the problems and potentials of decentralization.

The special problems associated with using renewable energy resources mostly stem from the intermittent source power flow. Fuel-based power systems are adapted to meeting fluctuations in power demand (load) because, within certain limits, the level of power production can be chosen at will. This is often impossible in the case of simple converters of renewable energy that do not incorporate energy storage facilities. Except for pumped hydro storage, energy storage is today considered too expensive to be practicable. If storage were economical, it would be used with present power systems to smooth the effect of the demand variations and hence allow constant production at the most economic power plants (base load units).

In the discussion of a possible transition to renewable energy sources, the regulation and management of the total supply system under conditions of variable power flow as well as demand must play a central role. It is therefore useful to distinguish two cases: one in which the total contribution of renewable sources into a given grid system is marginal, and the other in which such a contribution is substantial.

In the marginal mode there is, by definition, no change in the strategy of operating the fuel-based units. The presence of a small, intermittent, and fluctuating source of power input into the grid is absorbed in the regulation of fuel-based units, in the same way it presently absorbs load (demand) variations. Thus, in this mode, the decentralized input from renewable energy converters fully behaves as a negative demand. For this reason, an economic assessment should compare the full cost of the renewable energy conversion system with the marginal full cost of the fuel-based units. Owing to the marginal-mode assumption, the marginal fuel-based price times the renewable energy production at any given time should be integrated. The marginal fuel-based price is the price associated with producing the "last kWh" in a given utility system at a given time, a quantity monitored by utility companies.

If the contribution from renewable energy sources is more than marginal, the strategy of regulating the various units in the total system should probably be changed to reflect this in the optimal utility production plan. The characteristics of each renewable energy conversion system must be recognized in this procedure, which will be discussed in some detail below, in connection with examples of individual renewable energy converters.

GRID INTERFACING

When a fluctuating energy source is connected to an electricity grid, measures have to be taken to ensure stability and quality of power supplied to load points along the grid. Stability requirements include the absence of voltage excursions above a certain level at any of the load points, for example, when a particular renewable energy converter goes on and off the line. This problem can be serious if a large renewable energy installation is connected to the far end of a distribution line, with individual customers attached to the same line between the renewable energy converter and a transformer station.[2] Similar problems are not expected from the small units of the most decentralized systems, such as the rooftop solar cell systems. The inputs from large converters or from arrays of converters at the same location may have to be led to stronger points of the grid system or, alternatively, special switches and small storage facilities should be included in the system in order to obtain the desired suppression of too fast voltage excursions.

With respect to the "quality" of the power actually delivered to the grid from renewable energy converters, a variety of devices can assure a nicely shaped cosine alternating current (AC) of the prescribed frequency and under normal operating conditions can maintain the prescribed voltage. Some of these devices involve a feedback to the regulation of the renewable energy converter (for example, to the pitch control of a wind energy converter), while others take the output as it comes and modify it, using standard solid state techniques, to the specifications given. Some devices depend on model input from the grid to define the correct shape of the AC, but this feature can be eliminated so that the renewable power input need not be small compared with the total power flow in the grid.

If the grid receives input from many renewable energy converters situated in a dispersed fashion, the main parts of the grid are likely to receive an average input that is considerably smoothed relative to the individual inputs. For instance, the spatial distribution of short-time fluctuations in wind power may be regarded as random, implying that the main parts of a grid receiving wind converter inputs do not encounter any short-term fluctuations (cycle times below one minute). The grid sections near individual wind energy converters still encounter short-term fluctuations, of course.

Long-term fluctuations, that is, variations over periods of several hours or several days, may also be smoothed by having inputs from converters placed at a sufficient distance from each other that they encounter sufficiently different wind regimes. This should occur because wind is fairly site-specific and it is unlikely that wind would stop blowing at all sites on a grid at once—causing a more consistent average wind contribution. However, such fluctuations can be only partially removed by the dispersed siting method, and a reduction of seasonal variations would require the coupling of globally dispersed converters to a common grid.[3]

For direct collection of solar energy, the dispersed siting of individual converters may smooth out the short-term variations in power production occurring in situations of partially overcast skies, whereas the variation associated with the diurnal cycle can be smoothed only by a global network. The day-to-day variations in solar input, caused by local meteorological conditions on a synoptic scale, may become somewhat reduced by a sufficiently dispersed collector siting plan, but some problems similar to those described in relation to wind power still remain.

REGULATION OF MIXED SYSTEMS

For mixed systems containing decentralized, renewable energy inputs as well as fuel-based units, a new plan for regulation procedures has to be formulated. It is here assumed that the contribution of renewable sources has surpassed the marginal stage and that no energy storage facilities are attached to the grid under consideration. The need for a careful strategy derives from the prolonged start-up times of many fuel-based generating units. The most economical units, used for base load, typically have start-up times of several hours, while modern intermediate load units may have start-up times of about one hour (that is, the time from when a start-up decision is made to the time the unit is capable of delivering power to the grid). Only hydropower installations and special peak-load units (gas turbines and diesel engines) can be started within a few minutes.

Thus, if the contribution of renewable energy inputs to the grid falls short of expectations beyond the regulating capability of the fuel-based units already in operation, then peak-load units have to be started in order to make up the deficit. The peak-load units (for example, assuming that no hydropower is incorporated into the particular system considered) are characterized by high fuel costs, and the increased use of such units has a negative influence on the economy of the renewable energy converters. Therefore, modes of operation that reduce the amount of fuel spent in peak-load units are of paramount importance when we consider the introduction of renewable energy grid systems without existing power or energy storage facilities.

One example of a regulation procedure for the start-up of fuel-based backup units, which achieve the goal of supplying most of the auxilliary energy from base or intermediate load units, and negligibly little from peak-load units, has been described.[4] The idea is to rely on forecasts of production by the renewable energy converters when making advance decisions on the number of base and intermediate load units to start. Ideally, the forecasts may be based on meteorological theory, but it has been shown by straightforward simulation that by using extrapolations of wind power production based on previous records, at least 25 percent of the annual energy requirement for a specific system of wind energy converters in a fossil utility system can be provided by wind, without getting variability problems from incorrect forecasts, such as wasted surplus production or extended use of peak load units due to insufficient production. The precise percentage that can be handled by such reformed regulation may differ for different climatic regimes and different types of renewable energy sources, but the approach is generally applicable.

Special considerations must be made if some of the fuel-based units in the system are cogenerating (producing heat and electricity at the same time). At present such systems are normally operated in such a way that the coproducing units follow the heat demand load, and the lack of synchronization with the electricity demand load is made up for by imports and exports through the grid. If renewable energy units that produce only electricity with no storage are attached to the grid with cogenerators, the burden of additional regulation is placed on the cogenerators, which must operate following grid load and renewable energy variations, not just local heat considerations.

If a high percentage of the fuel-based units are bound to produce only because of the heat load, then it would be possible to add only a limited number of renewable energy converters to the system, because the "free" regulating capacity would be small. On the other hand, the introduction of other renewable energy converters designed to assist with heat supply may compensate for this. Such converters usually have a certain capacity of heat storage attached to them, and this opens up new possibilities for regulating the system.

With heat storage available, it is no longer necessary or desired to have the cogenerating power plants follow the heat load. Depending on which procedure is most economical, they may now follow the electricity load, or run at constant power level, or they may be regulated up and down according to the requirements set by renewable energy converters in the electricity system. In fact, only modest heat storage is required in order to obtain this freedom in system operation, and some cogenerating utilities are already today (without having renewable energy converters in their systems) installing heat storage facilities capable of accumulating the surplus heat produced during the day, if the generators are allowed to follow the electricity load. The heat is then circulated in the district heating network during the night, and the generators are required to cover the (low) night load of electricity so that at night only base load plants are started.

It is thus expected that with the inclusion of some heat storage, which may in any case be economically practicable, there would be no additional restrictions on the number of renewable energy converters that may be added to a mixed utility system. If most of the electricity generation were to be derived from renewable energy, heat pumps or renewable energy heat generators feeding into the district heat transmission lines would have to be added. That, however, would require storage facilities capable of delivering electricity, or access to adjustable renewable energy converters, such as hydropower plants or converters based on a storable harvest of biomass.

SYSTEMS WITH STORAGE FACILITIES

Large-scale energy storage with the possibility of recuperating electricity is now routine only for pumped storage in hydropower plants, an option restricted to certain geographical locations (otherwise underground reservoirs would have to be excavated, which currently is not economically feasible). Through grid linkage it may be possible for grid systems that do not possess pumped storage facilities to share the advantages of such installations with those utility systems that do.

Hydro installations with regulated flow from reservoirs to turbines may serve as energy storage facilities for other energy converters with intermittent output, even if water cannot be pumped upward. This is achieved by connecting the new energy converters and the hydro installations to the same grid. Then increased hydropower can satisfy demand if the other converters give a low output, while excess output from the new converters is used to replace hydropower. With proper rating, there will not be any net withdrawal from the hydro reservoirs associated with the introduction of the new converters.[5]

Efforts are being made to improve battery storage so that it may be used for utility purposes in large-scale systems or attached to small-scale systems. Several other storage concepts are now being researched, and it is quite likely that a number of short-term storage devices (capable of covering load in up to twenty-four hours) will be economically available within the next decade. Apart from hydro storage, the prospects for long-term storage are more uncertain. The uncertainty in making hydrogen storage cycles feasible, for example, is due not to technological problems but to economic ones. This means that if the energy resource situation demands that renewable energy sources be used, then 100 percent coverage is possible. The economic question is simply one of forestalling the price jumps of scarce fuels, and of government pricing and incentive policy.

Once energy storage is available, several alternative ways of operating an electricity system including renewable energy sources become available.[6] The general effect of storage is to make the intermittent resource behave more like conventional conversion plants. Depending on the capacity of storage, the system may be able to work at a constant level over increasing stretches of time; and with sufficient storage, the renewable energy system has outage times comparable to those of fuel-based units. Alternatively, the system may be operated so that it partially or fully follows the actual load, with the help of the storage facility. Finally, the renewable source may behave like a peak load unit if energy is (intermittently) accumulated in the storage facility during off-peak hours and is drawn from the facility only during peak-load periods.[7] Again, the particular renewable source, the amount of storage available, the load structure, and the cost of running different fuel-based units will determine the most economical mode of operation.

Ultimately, for an energy system using all renewable resources, the different types of renewable energy conversion and storage systems would have to be combined. For instance, part of the hydrogen that may be produced by surplus from electricity generators could be used for automotive or industrial processes, and high energy and density artificial fuels (made, for example, from hydrogen or from biomass) may in some cases be used to augment electricity production when short-term storage facilities are empty.

THE SUITABILITY OF INDIVIDUAL RENEWABLE ENERGY SOURCES

In the marginal mode the suitability of different grid-connected converters of intermittent energy sources may be judged simply by the average cost of energy production. This is not so if the contribution increases over the marginal level. Then the distribution of production over time will have to be considered, and also the structure of fluctuations in conjunction with adequate storage facilities to absorb such fluctuations and ensure reliable supply.

Solar radiation has a pronounced daily cycle and a latitude-dependent seasonal cycle. Dependence on cloud cover, turbidity (particle content) of air, and other factors also affect supply. Electricity demand also has a pattern of variation through the day, which may roughly follow the night-to-day and day-to-night changes in average insolation, but in most cases they are not identical.

Short-term storage or fuel-based backup facilities are therefore necessary in using direct solar conversion. Some schemes, such as running a thermodynamical engine on a fluid heated by (concentrating) solar collectors, may provide natural short-term storage, deriving from the possibility of storing the heated fluid from day to evening and night. No similar option is available for solar cells, which then require independent storage.

Biomass conversion schemes usually yield a storable fuel (gaseous or liquid), which makes them fairly free from intermittent supply and load problems. It is likely, however, that in the future liquid biomass fuels will be reserved for propulsive purposes, rather than for electricity production, owing to the scarceness of alternatives.

Hydropower often relies on water reservoirs, which make this source extremely flexible and perhaps even offer storage possibilities for other renewable sources. In many regions the reservoirs are at their lowest level when the average wind power is highest, so that this pair of renewable resources enhance the viability of the system above the level that either one of them could have achieved alone. Also, hydropower in off-peak periods can pump water for use during peak periods. For small-scale hydropower installations no storage options may be available, but, again, using them in combination with other renewable sources of a different seasonal variation may alleviate the need for long-term storage of energy.

Wind energy often has little pronounced diurnal variation, but at mid-latitudes it has a seasonal variation increasing with the height of collection. The highest average wind power is found in winter, which in many regions correlates well with the gross variations in electricity demand. This is not so in the southern United States because of the extended usage of electrically powered air conditioning devices. However, as the need for air conditioning correlates very well (obviously) with the availability of solar radiation, it would be reasonable to cover this load by direct solar conversion. Since the temperature differences aimed for are modest, it is not obvious that electricity is the best source of power for driving air conditioners. It would seem that simple flat-plate solar collectors, for example, combined with an absorption-type thermodynamical cycle, could provide the cooling and reduce peak electric load, and maybe even at a lower cost. The variability of wave power is similar to that of wind power, from which it actually derives its energy.

Fitting dispersed energy sources into the physical planning efforts of societies may place additional constraints on the composition of the system. The considerable space requirements of some renewable energy converters may pose a great obstacle to their acceptance if urban or arable land is to be bought for the purpose, or if constructions are proposed in recreational areas. Among the most space-consuming power converters are solar cell systems. If present low conversion efficiencies (about 10 percent for silicon crystal devices, less for cells made of amorphous material) prevail, large central power plants based on solar cells would require desert sites. A more promising solution, however, is to use existing rooftops, so that the converters do not have to bear the burden of separate land areas, separate supporting structures, and special transmission lines to remote sites. The modular nature of solar cell systems, with small solid state transformers and perhaps storage batteries attached to each array, makes them very flexible. Furthermore, the heat generated in the solar cells could be used for heating and hot water production. The regulation of a decentralized system using solar cells, including transmission of information on the decentralized production to the utilities' planning centers, which may have to furnish backup power, can be accomplished by means of microcomputers.

Flat-plate solar collectors for heating purposes are also most conveniently placed on rooftops or building surfaces, and competition between them and electricity-producing panels may arise if the available areas suited for either purpose are insufficient. The current attempt to develop hybrid units including both systems emphasizes the fact that renewable energy systems require societies to do their utmost to conserve energy, by insulation, proper system combination, and heat recovery devices, and by improving the efficiency of electricity-consuming equipment.

Biomass would normally be derived from the nonedible parts of plants, or from sewage or waste. Due to the priority of food production, energy "plantations" should be considered only at sea (algae or kelp) or in sparsely populated areas with marginal lands unsuited for raising food crops.

Wind energy converters have to be placed where there is suitable access for the wind. High windspeeds require extended "fetch distances," for example, over water, and such sites are found mostly along coasts but may also be found in extended plains or in certain mountain regions (the precise shape of mountain slopes is very important, and some mountain areas are disappointing from a wind-catching point of view). This means that acceptable wind energy sites tend to be concentrated in rather few locations. In such ideal sites many wind energy converters would have to be placed in arrays in order to arrive at adequate percentages of coverage of the total load of a grid system. The number of acceptable sites does increase with the height of the collecting device (tower). Still, many areas suitable for wind energy extraction would also be valuable for other purposes, which society may sometimes give higher priority to. For this reason it should be pointed out that if the land sites along a seashore are made unavailable, placing the wind energy converters at sea might be contemplated. If the water is sufficiently shallow, foundations to the bottom may be used, and if the shoreline has recreational value, the converters may be moved out far enough that no noise is heard on land and the converters appear similar to passing ships. In many cases the extra cost of going offshore may be at least partly compensated for by higher power levels usually found in the wind over the sea.

CAPACITY CREDITING

Most renewable energy sources flow intermittently, and it follows from the discussion above that it will not be possible for most systems without storage (hydropower again may be an exception) to guarantee that a certain load can be satisfied at a given time. For a mixed system consisting of a nonmarginal share of renewable energy converters as well as a number of fuel-based units, then, it may be said that there ought to be the same number of fuel-based units, as if no other converters were present, in order to be able to meet load at any time. The renewable energy converters should then be credited with only the fuel-savings yield, not the capital investment of other generators. However, fuel-based units are also not available all the time (due to maintenance and repair; in this connection only unplanned outage is relevant), and yet the utility attributes a capacity credit to each of them, according to certain rules. For instance, a small and isolated utility (in regard to its grid) may grant full capacity credit only to units with an unplanned outage rate of less than 10 percent of the time. It would then be reasonable to accord the same credit to a level of power production from renewable energy converters that can be assured for over 90 percent of the time. For most renewable energy systems, this level and hence the capacity credit granted would be considerably below the average power production. The capacity credit determines the fraction of capital costs that may be included in the economic evaluation of the renewable energy system. For instance, an array of wind energy converters may, during the year, deliver its average power, say, 50 percent of the time, and it may be capable of delivering 10 percent of the average power 90 percent of the time. If 90 percent of the time is the required availability factor, then in this case the wind energy system should get credit for 10 percent of its average power level, implying that the economical break-even point requires only that 90 percent of the cost should equal the price of the fuel displaced.

The question whether a partial credit should be given to some of the power available less than 90 percent of the time depends strongly on the structure of the utility system. If many peak load units are available, or if the grid system considered has strong links to other grid systems, with agreements providing for import at short notice without too heavy an economic penalty, then it is reasonable to demand less with respect to availability of individual plants, fuel-based or not. For example, the extensive grid connections among a number of European utility systems allow some of them to give full capacity credit to units available only 50 to 60 percent of the time.[8] This would imply that full capacity credit would be given to the wind energy converters in the example above.

If it is desired to increase the capacity credit of a given renewable energy system, for example, because fuel for back-up units becomes scarce or expensive, then energy storage must be included in the system. The fraction of time during which the combined system is unable to meet demand will decrease with increasing storage capacity and eventually will reach the level required for full capacity credit.[9] The amount of storage required to achieve this depends on the mix of renewable sources in the system and on the types of storage included, as well as on the variations in load and on the climatic regime in which the converters are operating.

ECONOMICS: METHODOLOGY AND PRELIMINARY CONCLUSIONS

An economic assessment of the viability of decentralized production units in electric utility grid systems may be conducted at different levels. We may concentrate on the economy for the utility company alone or for its customers, or we may take a general social point of view. Furthermore, the time scope of the economic evaluation may be shorter or longer and various indirect economic factors may be included.

A direct economic evaluation of energy-producing equipment can be made by estimating for each of a number of years (the "depreciation period") the cost of fuels, operation, and maintenance, and a fraction of the capital costs assigned to a particular year. The renewable energy systems do not have any fuel costs and, in general, different systems are characterized by differing emphases on the various components of cost. The distribution costs are not considered because the grid is here assumed to be the same, independent of the generating system (as discussed earlier, this may be only partially true if the sources are dispersed).

The capital costs are usually known when a given converter is installed, but the fraction of capital cost assigned to different years may derive from external rules (such as the annual installments of back-payment demanded on loans, if the capital is raised by a loan) involving such parameters as rate of interest, load duration, and load type. If the owner of the grid system (utility, state) covers the capital expenses out of its own assets, it may choose rather freely how to depreciate (that is, assign to different years) the sum paid. The common annuity-type depreciation or loan installments involve fixed annual costs. This means that the capital costs (plus interest accumulated) are evenly distributed over the depreciation period (which, in the most favorable case, would be taken as the physical lifetime of the converter), provided that there is no inflation. With inflation, annuity depreciation implies high annual cost during the first years, declining toward a lower cost during the last depreciation years, if quoted in fixed monetary values. In this case comparing the first year's cost will favor systems with the smallest capital costs, and an assessment aiming at comparing *long-range* viability should average the cost over all years, or should correct interest rates for inflation before making the evaluation.

The running costs, including operating, maintenance, repairs, and eventually fuel, can not usually be stated with certainty at the time of initiating operation. Future fuel prices are not known, and neither are the costs of labor and materials. Also, the amount of maintenance and repair that will be necessary is not often known in cases of first-generation converter types, for which no full lifetime experience is available. Although standard industrial methods exist for estimating such costs, some uncertainty deriving from the running costs must be considered in using the total cost figures, obtained by combining capital and running costs. The degree of government subsidy to some sources can also cause an imbalance in comparison.

Using this methodology of economic assessment—a depreciation time equal to physical lifetime and a fuel price for fuel-based alternatives rising as fast as inflation but no faster—it may be estimated that among the renewable energy converters now approaching commercial availability, some wind energy converters and some biomass converters, used in suitable environments, may already be economically acceptable for usage in modes not requiring separate storage facilities. Hydropower is already viable and in use in many places, and hydro systems may better allow larger contributions of other renewable energy sources (as discussed earlier) in connection with the storage and regulating options they hold. Photovoltaic conversion is not viable under the stated conditions until predicted dramatic extrapolations of cost reductions are realized. The same is true for systems comprising sizable energy storage (see chaps. 1 and 5) other than pumped hydro. The use of some short-term energy storage systems may already be close to economic viability.

Several factors in addition to clearly economic ones should be considered in planning for decentralization. They include the future outlook for scarce or expensive fossil fuels (for political or resource reasons), the increasing acceptability (safety and environmental) problems of nuclear conversion technologies, and the requirements for solving global development problems. These factors are very likely to include major changes in life-style in today's most developed countries. One of the key words in present discussions of life-style changes is "decentralization," and although the importance of this concept lies primarily in questions of human relations and social structure, it is very probable that a decentralization of energy production may both stimulate the transition and be required by it. Still, just as there must be communication and information flow among the parts of a decentralized society, there is also a clear advantage in maintaining energy transmission systems such as electricity grids. They serve to exchange power in the case of load mismatch, and they may serve as a security in case of failure or other problems of individual conversion units. There is a real need for objectively establishing the most efficient degrees of decentralization, taking into account the dependence of system performance on size and siting of individual converters and energy storage facilities.

NOTES

1. Connections between choice of energy technology and development goals are exploited in B. Sørensen, "Energy Technology and Social Structure," Proc. Int. Symp. on Technology Appropriate to Underdeveloped Countries, UCA, San Salvador, El Salvador, February 1979; *New Scientist*, August 16, 1979, pp. 513–15.

2. Cf. G. Elsoe Jorgensen, M. Lotker, R. Meier, and D. Brierley, "Design, Economic and System Considerations of Large Wind-driven Generators," Proc. Institute of Electrical and Electronic Engineers Winter Meeting of Power Engineering Society, 1976.

3. Investigations have been made by J. Molly, *Wind Engineering* 1 (1977):57–66, and E. Kahn, Lawrence Berkeley Laboratory Report LBL–6889 (1978).

4. B. Sørensen, in Proc. 2nd Int. Symp. on Wind Energy Systems, Amsterdam 1978, vol. 1, paper G1 (British Hydromechanics Research Association Fluid Engineering, Cranfield, UK, 1978). See also n. 9.

5. B. Sørensen, *Energy Policy*, March 1981, pp. 51–55.

6. J. Haslett and M. Diesendorf, *Solar Energy* 26 (1981):391–401.

7. B. Sørensen, *Solar Energy* 20 (1978):321–31; W. Coste and M. Lotker, *Power Eng.*, May 1977, pp. 48–51; S. Leonard, Proc. 12th IEEE Photovoltaic Specialists Conf., Baton Rouge, 1976, pp. 641–52.

8. See, e.g., M. Johansson, *Danish Electricity Supply Undertakings*, Report DEFU–TR–152 (1974), English translation by NASA: TT–F–16058.

9. Further details, and a discussion of indirect economic factors, may be found in B. Sørensen, *Renewable Energy* (Academic Press, December 1979).

ROBERT E. WITHOLDER, JR.

CHAPTER 7 | *Economic Feasibility of Dispersed Solar Electric Technologies*

Most of the contributors to this volume mention the economics of alternative electricity systems as one important factor in the planning process. Pricing the possibilities, however, is difficult at best in the energy field. Brown and Lovins both assert that our values as a society largely determine economic feasibility. Further, the economics and sensibility of alternative electrical generation technologies vary according to geography and other local conditions.

Witholder tackles the economic problem in this chapter, a summary of a multi-year study by the Solar Energy Research Institute (SERI) of the economics of various solar technologies in different locations. He claims that it is not possible to determine economic feasibility by studying a single variable or by using the same measures in all regions. Regional environmental conditions, local utility profiles, and the future costs of the technologies are all part of the analysis. He presents an analysis of technologies in several locations and concludes that, using the defined methodology, wind systems are already economically feasible in certain areas while in others photovoltaics may not be economical well into the twenty-first century. Perhaps as important as Witholder's conclusions is his explanation of the methodology used in the SERI study for assessing such feasibilities.

This article summarizes the findings of regional assessment studies conducted by the Solar Energy Research Institute in Golden, Colorado, of the market feasibility of various distributed (user-owned) solar electric technologies. Regional assessments of the technologies were undertaken because three of the four critical variables in determining feasibility are region specific. The four variables are: the available solar resource, regional market economics, utility characteristics (reliability and production cost), and the projected future cost of these technologies. Thus, unless they are based on specific regional studies, generalizations are simply not useful. The distributed solar electric technologies studied were: (1) wind turbines; (2) photovoltaics; (3) solar thermal facilities (primarily dishes, but the data can be related to parabolic troughs, distributed line focus systems, and solar ponds by making linear adjustments for performance and cost). Cogeneration, which has similarities to the other intermittent technologies, was not analyzed; however, the key issues of sellback of excess energy to the utility and impact on the utility are directly applicable to it.

The economic feasibility of a distributed solar electric technology is established when the cost of the system is equal to or less than what the user can afford to pay (that is, the value of the system to the user). In the initial 10 percent penetration of a market, almost any innovation is governed principally by buyer or user behavior, which is generally independent of the economic value of the system. For example, the user may want the system for other than economic reasons. This was the case in the distributed solar energy market from the late 1970s to 1980. With the exception of wood burning and new building construction incorporating solar thermal–hot water applications, passive concepts, and small wind turbines, few solar technologies at the start of the energy crisis passed the rigid test of economic feasibility. This is especially true of the distributed solar *electric* technologies (wind turbines, photovoltaic systems, and solar thermal electric systems), and the reasons are discussed below. This situation is not likely to continue in the future because of rapid improvements in performance and cost reduction of solar *thermal* technologies and because of equally rapid increases in conventional fuels. When the cost of the system to the user is equal to what the user can afford or to the money-saving value of the system, further penetration can take place. Thus the market is usually governed by relative economics.

The findings of the analysis of dispersed applications can be summarized as follows:

1. Provided the wind resource is available, small and intermediate wind machines have in general the best economic feasibility of the solar electric technologies.
2. Photovoltaic systems have the next highest potential and in some regions (such as the south central United States) may have higher potential than wind turbines because of the availability of the solar resource.
3. Solar thermal dispersed generation of electricity may have the least potential because of the lower availability of the resource and the cost of the system. Solar thermal can have the highest potential provided it produces thermal as well as electrical energy (total energy) to satisfy dispersed industrial demand.

COST VERSUS VALUE

The two factors that must be taken into account in determining the economic feasibility of solar electric technologies are: (1) the *cost* of the system and (2) the *value* of the system to the user, or what the user can afford to pay for the system. Figure 1 shows the simple relationship of cost and value as a function of time.

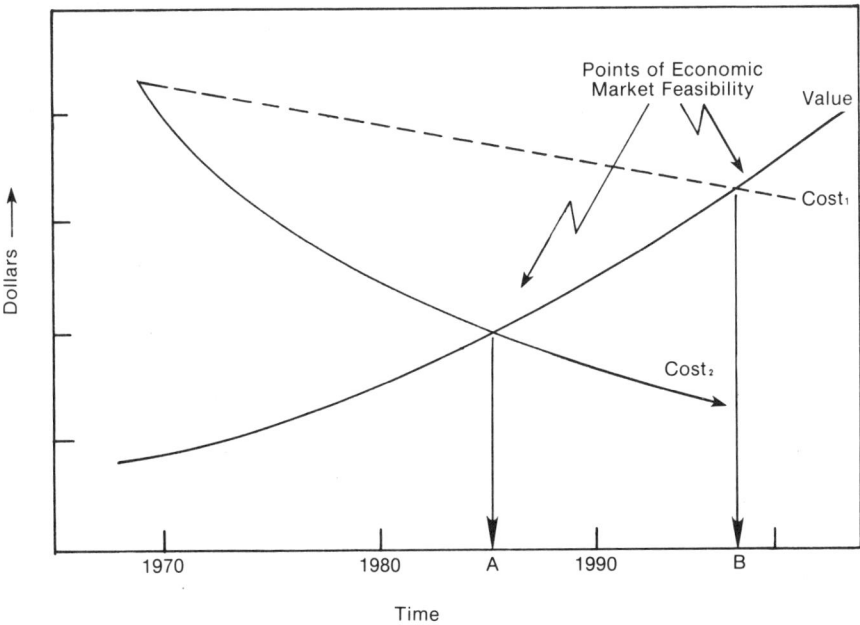

FIGURE 1. Simplistic Relationship of Cost versus Value for Distributed Solar Electric Technologies.

Two cost curves are shown in figure 1. Cost 1 represents the hypothetical cost of a technology for distributed solar electric application that is decreasing in time (owing to innovations in the technology and mass production). Cost 2 is decreasing even faster than cost 1. Cost 1 would have been the situation for most of the solar electric technologies (photovoltaics, solar thermal, and intermediate to small wind turbines) without government intervention and privately funded research and development. Eventually these technologies would become economically feasible (point B), probably at the turn of the century (assuming little government R & D or incentives), because of the action of market forces independent of the technology (for example, rising conventional fuel costs that translate into higher electricity cost). What has happened instead is more in line with cost 2. The economic feasibility of the solar technologies has been accelerated to an earlier time (point A) by a modest investment (tens of billions of dollars) of research and development funds (as compared with the hundreds of billions of dollars invested in new systems that will be installed), coordination of activities, information dissemination, and a relaxation of institutional barriers (solar access rights, and insurance and regulatory changes including rate adjustments and sellback).

As the result of intensive analysis of the systems (from the point of view of innovation and suitability for mass production), solar technology developers can project future costs and performance as a function of time. Future costs are shown in tables 1, 2, and 3, but they represent only the capital investment in an installed system (FOB). A user or decision maker could be misled if he com-

TABLE 1. Projected Cost of Distributed Photovoltaics

Distributed Photovoltaics[a]	1982	1985–86	1990–2000
Module cost (FOB)	$2.80/$w_p$	$.70/$w_p$	$.40–.15/$w_p$
Installed System Cost	$6–13/$w_p$	$2.60–1.60/$w_p$	

a. National Photovoltaics Program; DOE, Sept 1980, Jet Propulsion Laboratory

pared these capital costs with the conventional system capital costs (or with each other) in making a purchase decision. For example, distributed solar electric systems are capital intensive (yet have no fuel cost component) when compared with conventional electric sources.

The user of conventionally generated electricity (either utility or distributed) will have to pay only a little or no initial cost (perhaps a hook-up charge and a heater) and a monthly bill, which will continue to rise as long as fuel costs increase. On the other hand, the would-be owner of a distributed solar electric system may be confronted with a high initial cost (down payment and/or mortgage payments) when he purchases the system and the thought that it may take up to twenty years to recover the cost in energy savings. For new construction—residential, commercial, or industrial—this cost is somewhat mitigated since the mortgage of the plan (or residence) can include the system with a payment for it spread out over thirty years. For retrofits or industrial repowering (the larger market when compared with new construction) this is not the case. The initial financial burden will be two or three times higher than the price the owner would have paid on a monthly basis for energy from a utility.

Numerous approaches are being suggested to eliminate the initial high cost of purchasing a distributed solar technology, for example, guaranteed loans, utility ownership, third-party ownership, various leasing arrangements, and tax incentives. In the SERI regional assessment studies it was assumed that the payback on investment over a reasonable time interval is critical to the decision to purchase distributed solar electric technologies and that the barrier of initial high cost can be removed from their deployment.

The second factor in determining market feasibility is the value of a system to a user, or what the user can afford to pay for it. To determine value, we must first determine what the user would pay if he purchased all his energy from the

TABLE 2. Projected Cost of Distributed Solar Thermal System

Distributed Solar Thermal Systems[a]	1982	1985	1990	2000
Line-Focus Systems[a]	$206/m²	$160/m²	$96/m²	—
Point-Focus Systems[a]		$450/m²	$120–170/m²	$80–120/m²
Installed System Cost ($/kWe)	—	*$4200/kWe	*$1300–1900/kWe	*$800–1100/m²

a. R. B. Edelstein, *Solar Thermal Cost Goals*, ASME/Solar Conference, May 1981: SERI/TP-633-1063.

TABLE 3. Projected Cost of Small and Intermediate Wind Systems

Wind Systems			Installed Cost ($/kWe)		
Application	Model	Manufacturer/Sponsor (size)	1980	1990	2000
Single Family	1500	ENERTECH (1.5 kWe)[a]	3200	1620	1020
Residence	SI4	Dakota Wind and Sun[a] (4 kWe)	2400	1750	1000
Low-Rise Apartment	SI20	Dakota Wind and Sun[a] (20 kWe)	1570	1250	1000
Farm	SI20	Dakota Wind and Sun[a] (200 KWe)	1570	1250	1000
Shopping Center	MOD-OA	DOE (200 kWe)[b]	1800	1140	1000
Fluid Milk Plant	MOD-OA	DOE[4]	1800	1140	1000
Pulp and Paper Mill and Utility	MOD-2	DOE[4]	1600	860	860

Costs do not include land, utility interface equipment, and annual recurring cost.
a. These costs were compiled by contacting the manufacturers.
b. Northwest Regional Assessment Study, Preliminary Draft, SERI, March, 1981.

electric utility over the lifetime of the solar technology (or over a suitable period of time to recover his investment), using appropriate assumptions about likely cost escalation. We must also determine the cost of the same amount of electricity from the grid. It is also important to consider the value of any excess electricity generated by the solar technology that the user is likely to sell to the utility for credit against his utility bill.[1]

By subtracting the cost of energy using solar technology from the cost of energy without solar technology, the savings in energy costs attributable to the solar technology can be determined. The savings over the time required to recover an investment—the value of the system—can now be compared with the cost of the system. If the value (in dollars saved) is greater than the cost of the system, it is clear that the purchase of the system should be seriously considered. By dividing the annual value (yearly savings) into the life-cycle cost, we may derive an approximation of investment payback in years that can be compared with a similar approximation for other choices (either solar or conventional) to give a relative ranking of the choices. The system with the lowest number should be chosen. The determinations of value vary widely in different regions of the United States, causing national generalizations to be inadequate.

REGIONAL DETERMINANTS OF VALUE

Three principal regional variables significantly influence the value of a given distributed solar electric technology. They are, in order of importance: (1) the solar resource available; (2) the type of utility (and, indirectly, the cost of fuel used by the utility); and (3) the correlation (or match) in time among the solar energy output, the electrical demand of the user, and, to a lesser extent, the electrical demand placed upon the utility. The first two characteristics obviously affect the value of solar energy to the user, while the third concerns the unique character of distributed solar electric (and possibly cogeneration) systems because it determines the amount and sellback rate of excess energy (electrical energy not used by owners of the solar technology) that can be sold to the utility grid.

METHODOLOGY FOR DETERMINING VALUE

The methodology that was used to derive the value of various distributed solar electric systems is illustrated in figure 2. It involves five analytical steps:

1. *Solar Technology Performance:* Utilizing hourly[2] solar resource data for a site as an input, hourly estimates of the power output from the solar technology are prepared. Performance analysis models for photovoltaics, solar thermal electric, and wind energy conversion systems are used in this step.
2. *Energy Use Model:* The hourly output of the solar technology is compared with the user's load to determine the degree of excess or shortage of solar-generated electricity. Shortages require backup energy from the utility grid, while excess is sold to the utility at sellback rates.
3. *Utility Simulation:* Hourly electric power output profiles are scaled to reflect the assumed capacity of solar electric systems in place. The hourly load profile for the utility is then adjusted to reflect the effects of the hourly excess electric outputs of the solar technologies. The output is a set of load duration curves, consisting of the baseline load duration curves (without solar) for each subperiod, and solar adjusted load duration curves representing the effect of the excess solar electric and solar energy consumed by the user of solar electric energy on the load served by the conventional portion of the utility.

   A probabilistic utility production costing routine can be used to provide a refined estimate of production costs for both the baseline load duration curve and the load curve with the capacity mix adjusted to reflect the addition of the solar electric technologies. The input loss-of-load probability, maintenance schedule, fuel costs (see table 4), operating strategy, and generating capacity mix are used as the basis for these calculations. The output of this step or model is a rate charged for energy sold to the user and a sellback rate for energy sold back to the utility.[3]
4. *Savings Analysis:* The outputs from steps 2 and 3 are used to provide estimates of the value of the solar electric system being analyzed. The user's saving is derived from (1) the cost of energy that the solar technology displaces (the load times the utility rate) and (2) the energy in excess of the load times the sellback rate. The sellback rate may be set below the utility savings in generating cost (for example, 25 percent less) to allow for additional metering and administrative expenses incurred in the purchasing of excess energy from dispersed systems. Estimates of the savings

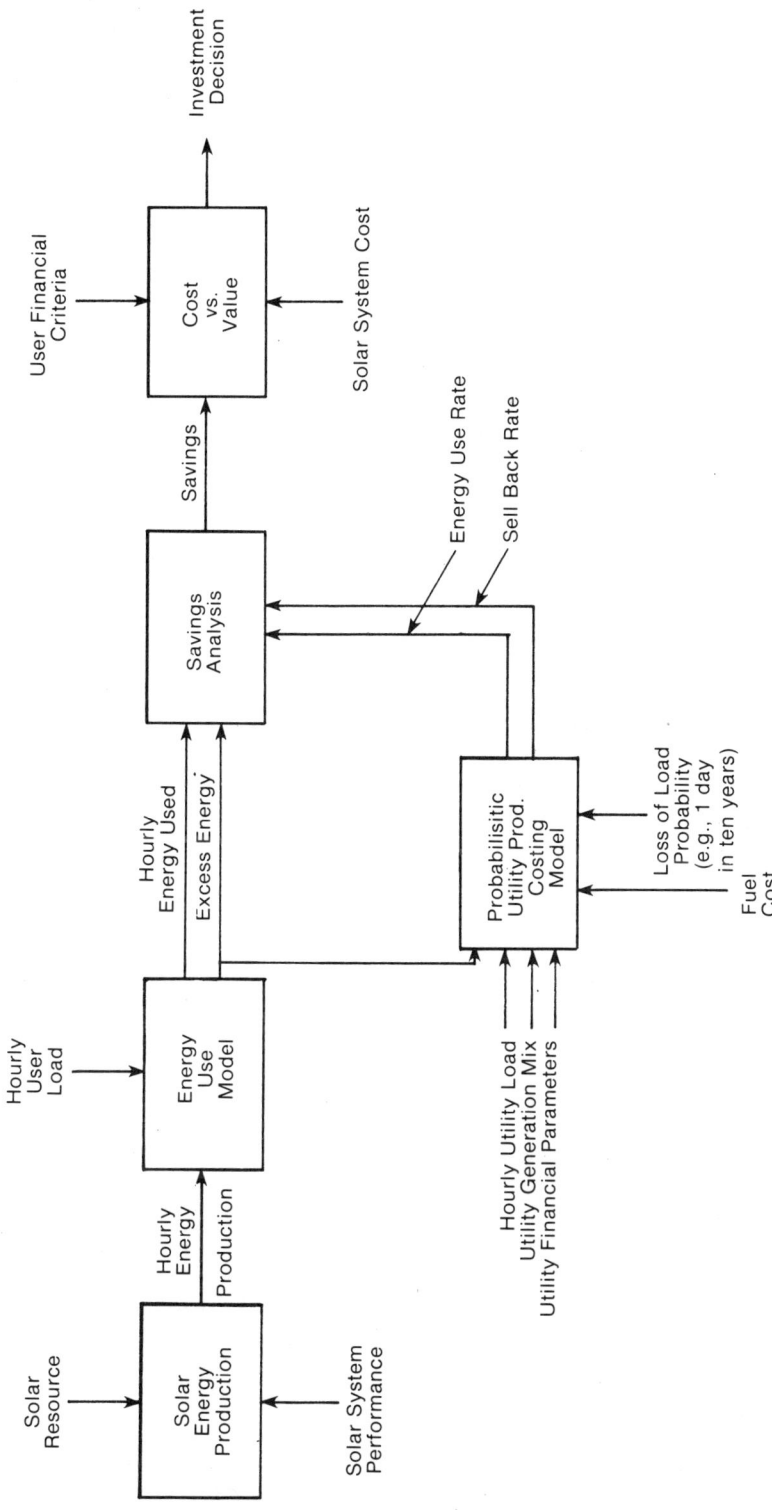

FIGURE 2. Value Analysis for Distributed Solar Electric Systems.

TABLE 4. Baseline[a] Fuel Cost ($1980/MMBtu) Used

	Region					
	South central		North central		Northwest	
Fuel Type	Cost	Escalation (%)	Cost	Escalation (%)	Cost[b]	Escalation (%)[b]
Nuclear	.70	2	TBD	3%	1.8	5.9–9.5
Coal	1.40	2	TBD	3%	4.21	8.4–9.9
Residual Oil	3.10	3	TBD	3%	2.66–3.3	6.3–10
Distillate Oil	4.50	2.5	TBD	3%	4.15	6.3–10.35
Natural Gas	2.30	4	TBD	3%	.95	6.3–7.5

a. A range of cost and escalation factors were used in determining the sensitivity to these costs.
b. Depends upon utility in the region.

are translated into a value of the solar electric technology on the basis of the user's purchase decision criteria.

5. *Cost versus Value:* The value of the solar electric system to the user is compared with the cost of the system. For example, the value can be accumulated on an annual basis for either the lifetime of the system (20 to 30 years) or for the number of years that the investor would want to recover his cost (for example, 5 to 7 years). Then, using an appropriate set of economic constraints specific to each user (see table 5), the break-even capital (installed) cost[4] of the system can be determined. It can then be compared with the capital cost of solar electric systems, or value–cost ratios can be compared.

TABLE 5. Distributed User Economic Parameters for the Northwest, North Central and South Central Regions (Different assumptions are in parentheses)

Economic Parameter	Residential	Agricultural	Commercial/Industrial
Down Payment (%)	20 (15)	0 (25)	0
Interest Rate (%)	10 (9)	10	10
Loan Term	20 years (30)	20 years (10)	20 years (30)
General Inflation Rate (%)	6	6	6
Depreciation			
—Variable	N/A	2 (double-declining)	2 (double-declining)
—Term	N/A	14 years (10)	14 years (10)
Annual O&M (% of Capital Cost)	2[a]	2[a]	2[a]
Tax Credit (%)	0	10 (investment) (20–40)	10 (investment) (20–40)
Sales Tax Rate (%)	4	4	4
Property Tax Rate (%)	2	1.5 (2)	2
Marginal Tax Rate (%)	35 (30)	25 (50)	48 (50)
Annual Insurance (% of Capital Cost)	0.5 (.25)	0.5 (.25)	0.5 (.25)

a. In the south central region a range of O & M costs were used for each technology that declined in some cases as the technology matured.

THE SOLAR RESOURCE

The availability of solar energy is marked by wide geographic variations in intensity and usefulness as well as in the more commonly recognized changes with time of day and season. A characterization of the insolation and wind resources is required for each particular solar electric technology because the output of each is affected by the solar resource availability at the site.

INSOLATION RESOURCES. Because valid long-term insolation resources are measured at relatively few places, geographic interpolation is required to estimate the solar resource at most other sites. A sample of eight sites, shown in figure 3, was selected to provide the insolation data required to determine the output from photovoltaic and solar thermal systems.

The output of a photovoltaic system is directly dependent on the amount of light incident on the photovoltaic cells. For flat plate systems, the relevant insolation is the amount of insolation incident on a surface of panels tilted to the latitude angle (both the direct and diffuse radiation from the sun). For concentrating systems that use photovoltaics and solar thermal technologies, the insolation is the amount of energy incident on a sun tracking surface or aperture (the "direct normal" insolation as shown in figure 3).

Hourly data on insolation are available only at a few SOLMET (solar meteorological data base) sites in the regions studied (cities shown in figure 3). These data were used in the value analyses for this study.

WIND ENERGY RESOURCE. The wind resource is measured from wind speed indicators at fixed heights above local ground level. Wind energy is proportional to an exponential power of wind speed; thus particular care must be taken in averaging and integrating. Due to boundary layer effects, the measured wind speed at the instrument height has to be extrapolated to the (usually higher) hub height of a wind machine.

The wind resource is significantly affected by the terrain. The variation in the power available from an instrument site to a "good" wind site has been reported to be as high as a factor of two to four. Based on national wind energy assessments, the map of figure 4 presents the long-term average wind power (in W/m^2). Also shown in figure 4 are the sites used in this study. The contours shown are for a level above the surface of approximately 50 m, corresponding roughly to the hub height of megawatt-size wind turbines. Appropriate adjustments were made for smaller wind turbines in accordance with the appropriate power law for wind-speed changes with height above the ground. The results are believed to be accurate within 25 to 30 percent, with the main inaccuracies due to the extrapolation procedure and to the relatively low density of data stations. The wind-power values presented are thought to be conservative, with higher values expected in hilly and coastal areas.

The wind power available is greatest in the winter and spring in most of the United States. Large regions of northern Texas, Oklahoma, and especially Kansas have power densities of 400 to 500 W/m^2, which are among the highest in the country. High power areas are also expected along the coastlines.

Proper siting of wind turbines requires knowledge of the available wind energy at the specific sites for a minimum of three years. This could be accom-

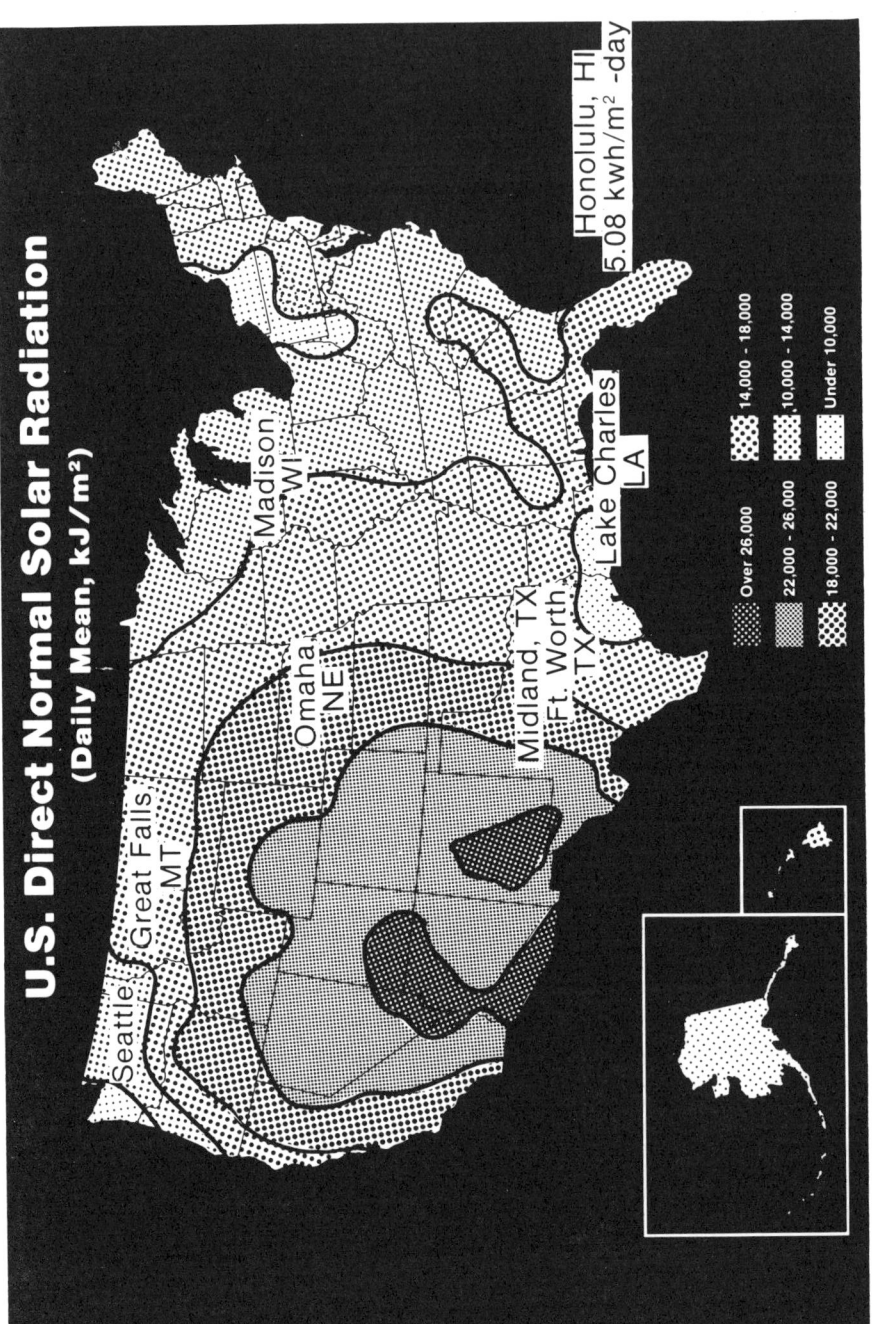

FIGURE 3. The Sites Investigated Using Direct Normal Solar Radiation (kWh/m²-day). Source: Data Collected by the DOE National Insolation Resource Assessment Program, Solar Energy Research Institute, head center.

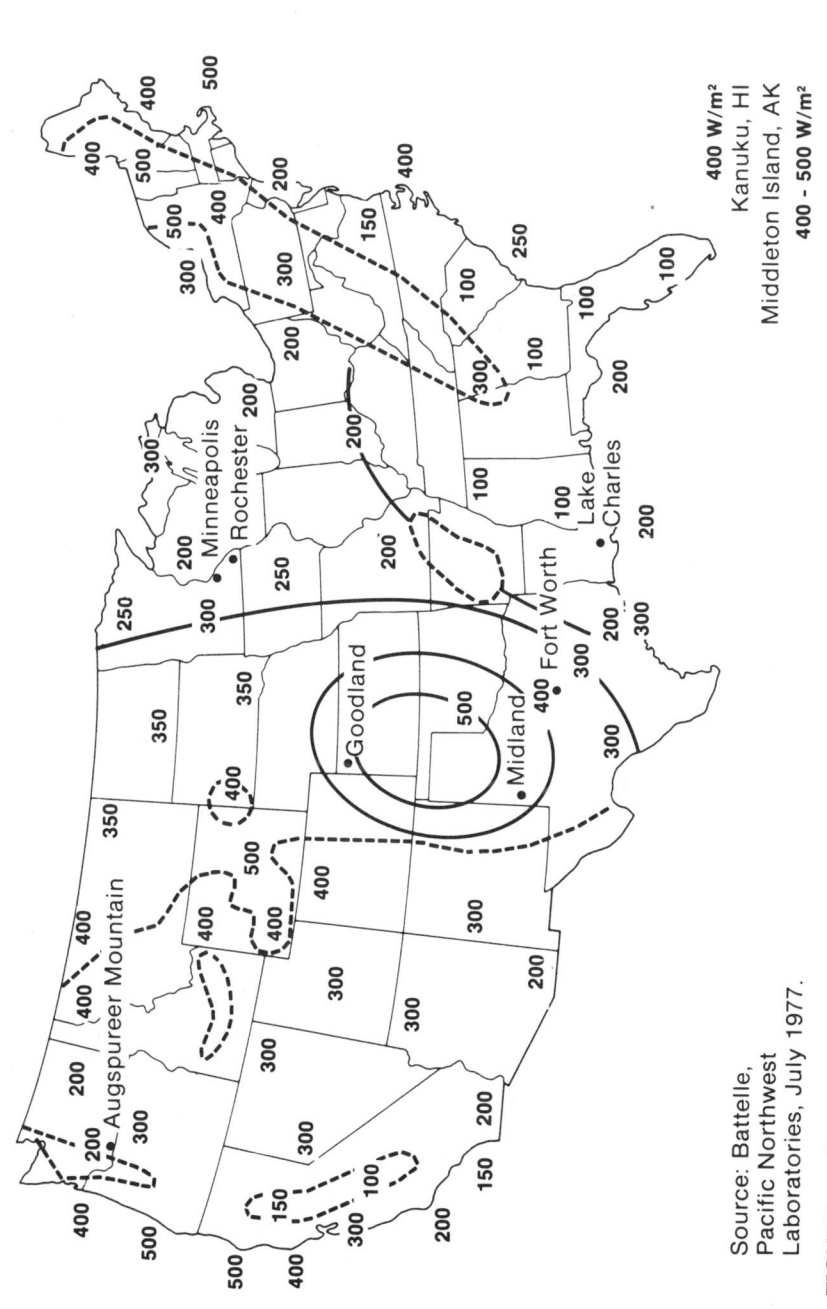

FIGURE 4. The Wind Sites Investigated (W per m²).

Source: Battelle, Pacific Northwest Laboratories, July 1977.

plished by installing anemometers and other test equipment at the proposed site, and by taking measurements over periods of time. The time variability of wind energy makes tests of shorter duration unrealistic predictors of energy availability.

Specific wind resource values for sites shown in figure 4 are provided by the SOLMET data. Wind power densities at these locations range from 161 W/m^2 to 556 W/m^2.

Utility Type

The type and characteristics of a utility in the user's locality are critical to determining the economic feasibility of distributed solar electric technologies. The value of a solar electric technology energy is a function of the rate charged for conventional electricity and the rate for sellback of excess energy to the utility. Both rates are a function of the type of utility. The two primary characteristics that determine these rates are: the utility ownership and the utility generation mix (which determines the fuel usage and consequently the greater part of the utility's avoided cost when the utility's energy is displaced by a distributed source).

Utility Financing Alternatives

There are, in general, four major utility ownership alternatives. These are municipal utilities, rural electric cooperatives (RECs), investor-owned utilities, and federal and state power authorities. Approximately 77 percent of the nation's electricity requirements are met by investor-owned utilities. Of the remaining 23 percent, over half is generated by municipal utilities and rural electric cooperatives. Federal power administrations produce the rest. As shown in table 6, the generating mix capacity varies substantially among utility types. For example, rural electric cooperatives obtain over 80 percent of their energy requirements from coal, while federal power administrations rely primarily on hydroelectric power. The Tennessee Valley Authority, a major federal power administration, also has nuclear capacity. Municipal utilities are projected to continue using greater than average amounts of oil for power generation. The principal differences among these alternatives are in financial structure and size. The four alternatives are ranked below in the order of their attractiveness to solar economics.

MUNICIPAL UTILITIES. Municipal systems are typically small (200 MWe or less) and generally serve only a single municipality or metropolitan area. Municipal systems are usually administered by a board of directors and in some states come under the jurisdiction of the state utility commissions.

Some power may be furnished by generating equipment owned by the municipal system, but there is a strong trend for small municipal systems to buy power from investor-owned systems because of the high cost of generating electric power using combustion turbines and generating units driven by internal combustion engines.

TABLE 6. Summary of Generation by Fuel Type and Utility Ownership Type for 1978 and 1990

	Coal (10⁶ MWh)	%	Oil (10⁶ MWh)	%	Gas (10⁶ MWh)	%	Nuclear (10⁶ MWh)	%	Hydro (10⁶ MWh)	%	Total (10⁶ MWh)	%
1978												
IOU	801	47[b]	338	19[b]	273	16[b]	230	14[b]	73	4[b]	1715	77[c]
MUNI	49	23	36	17	26	12	28	13	77	35	216	10
REC	48	83	2	3	6	11	2	3	0	0	58	3
FPA	78	34	2	1	0	0	17	8	130	57	227	10
TOTAL	976	44	378	17	305	14	277	12	280	13	2216	
1990[a]												
IOU	1640	59	295	11	60	2	690	25	75	3	2760	77
MUNI	170	45	48	13	9	2	65	17	82	22	374	10
REC	68	83	2	3	3	4	8	10	0	0	81	2
FPA	93	26	0	0	0	0	120	34	139	39	352	10
TOTAL	1971	55	345	10	72	2	883	25	296	8	3565	

IOU = Investor-owned utility REC = Rural electric cooperative
MUNI = Municipal utility FPA = Federal Power Administration

a. Source: General Research Corporation projections based on 1979 Reliability Council data, August 1980. Nuclear totals for 1990 adjusted down by SAI to reflect energy demand growth of 4.1% per year.
b. Percent of electrical energy for specified utility type provided by designated fuel.
c. Percent of total electrical energy provided by designated utility type.

Municipals are financed by the issuance of tax-free municipal bonds. The attractiveness of these bonds in the market is generally related to the overall financial condition of the municipality. The potential access to low-cost capital and the municipals' tax-free status make municipal utilities an attractive potential owner of solar electric systems.

RURAL ELECTRIC COOPERATIVES. RECs are systems financed and regulated (to some extent) by the Rural Electrification Administration (REA) of the U.S. Department of Agriculture. RECs serve rural areas almost exclusively and range from very small (5 MWe) transmission and distribution (T&D) co-ops to fairly large (100 to 1,000 MWe) generation and transmission (G&T) co-ops (called member cooperatives), which in turn serve rural customers.

RECs obtain loans through the REA at low to moderate interest rates depending on the size of the REC. The relatively low cost of money compared with direct funds for RECs has prompted the RECs to participate in capital-intensive generation projects (such as nuclear plants). RECs are also regulated by the state utility commissions.

FEDERAL AND STATE POWER AUTHORITIES. Federal and state power authorities are generally bodies set up to administer and operate small to medium-size (100 to 1,000 MWe) federally or state-financed power projects such as hydroelectric installations. Power from these sources is usually sold to other utilities in the region under terms specified by the power administration and approved by the state utility commission.

INVESTOR-OWNED UTILITIES. Investor-owned utilities range in size from 200 MWe or less up to very large systems of 12,000 MWe. These utilities typically serve urban centers and some of the surrounding suburbs, and they sell power to small municipals and RECs for distribution to their customers.

Investor-owned utilities raise capital by issuing stocks and bonds and must depend on prevailing interest rates. In addition, they are taxed as all other profit-making corporations in the United States. The attractiveness of a utility's stocks and bonds in the market depends on its general financial condition.

The regions most favorable for distributed solar electric deployment from the perspective of the utility are those served by municipal utilities (owing to their tax-free status and low cost of capital) or by utilities that have a high dependence on oil or natural gas, which will be the highest-priced fuels. Also, municipal utilities tend to be shorter on capital and thus more open to distributed generation. There is no geographic distribution or correlation with utility financial structure, but figure 5 shows the projection of oil and gas usage for 1978 and 1990. Historically, oil and gas have been the primary fuels for electric power generation in California, the south central region, Florida, Hawaii, Puerto Rico, the U.S. island territories, and portions of the Northwest. However, as new coal and nuclear capacities expand, oil and gas use is expected to decrease substan-

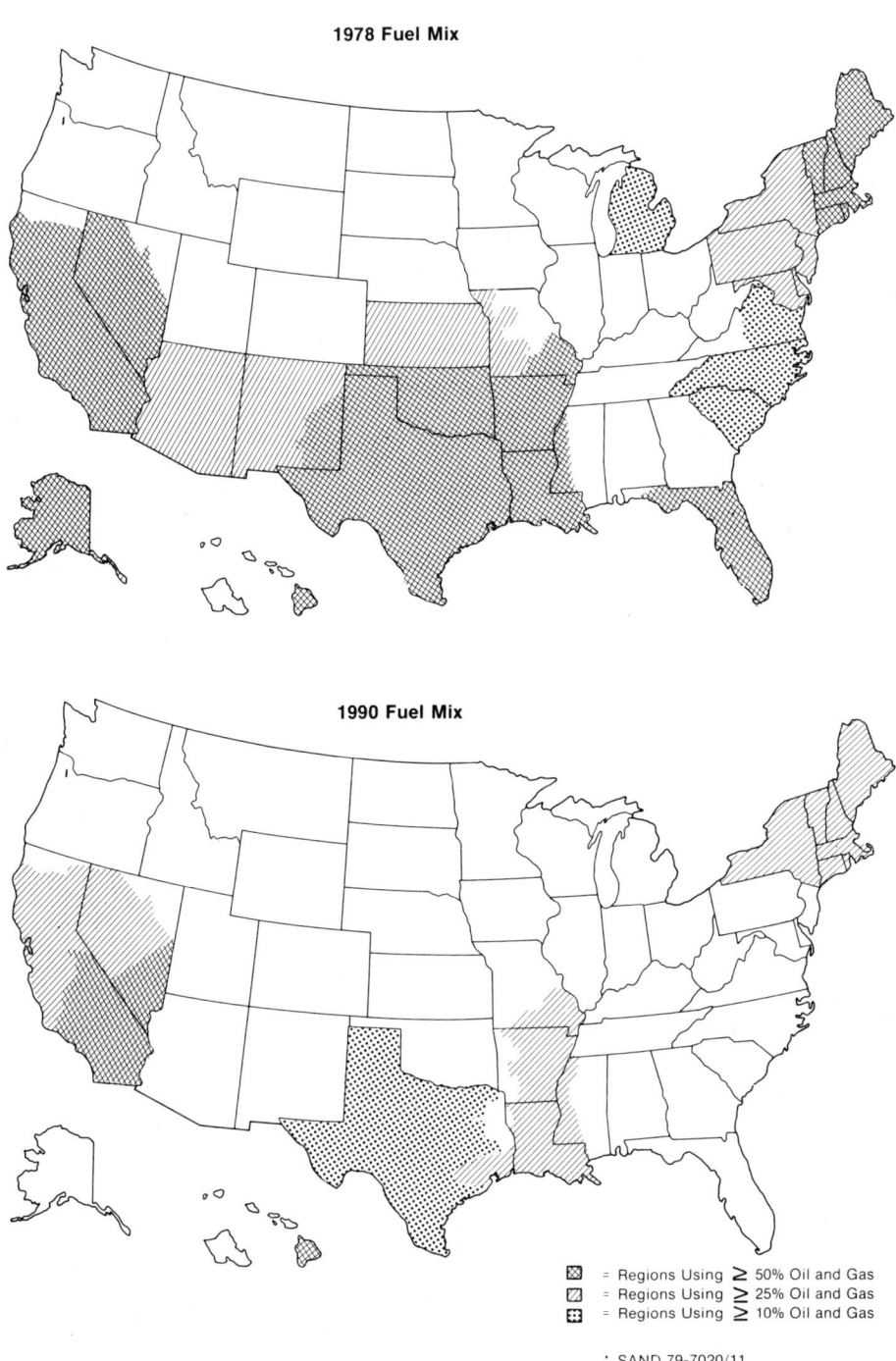

FIGURE 5. Oil and Gas Generation—1978 and 1990.

tially in most regions. By 1990 Puerto Rico, Hawaii, and the Southern California/ Nevada region are expected to use oil for more than 50 percent of their power needs, and the Northeast for somewhat less than 50 percent of its needs.

Output Versus Load

Unlike conventional energy sources, solar energy is intermittent. As a result, its availability will not necessarily follow the load. The two technical alternatives for most distributed users are: (1) to utilize the utility grid when energy is not available from the solar technology and to sell energy back to the utility when

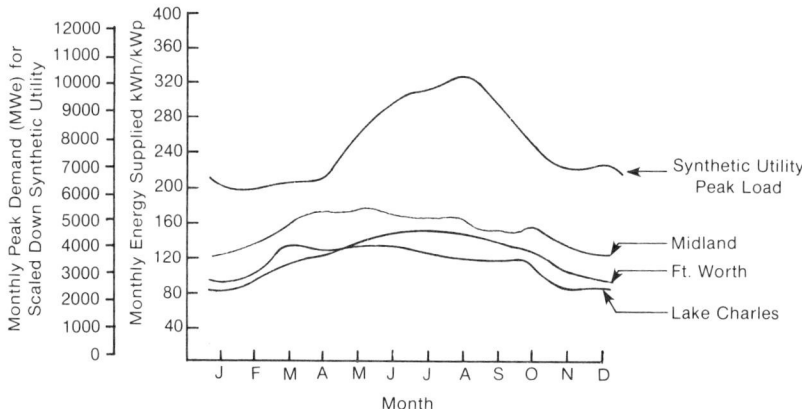

Monthly Energy Supplied by 10 kWp Residential Photovoltaic System

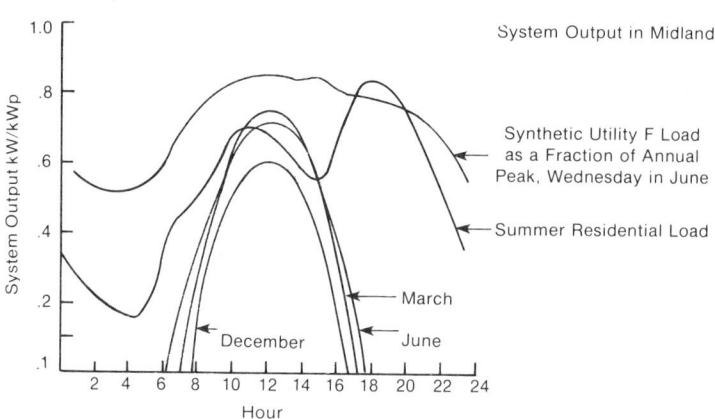

Average Daily System Ouput — 10 kWp Residential Photovoltaic System

FIGURE 6. Flat Panel Residential Photovoltaic System.

THE ECONOMICS OF DISPERSED TECHNOLOGIES 137

too much is available from the technology; and (2) to provide storage for the solar energy. An economic analysis of battery cost versus use of the utility grid (at utility rate) shows that the utility grid is the proper choice as backup. Storage will become a practical consideration (or necessity) for distributed sources when the total capacity of distributed solar sources connected on a given grid exceeds approximately 20 percent of the capacity of that particular utility's peak conventional capacity.

Assuming no storage, the normalized output of various dispersed solar electric systems for the south central region of the United States is shown in figures 6 through 10. Also shown are the typical normalized loads for a residential and

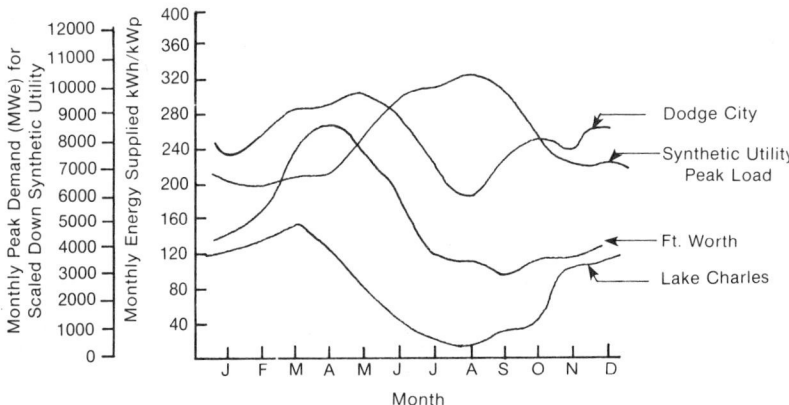

Monthly Energy Supplied by a 10 kWe HAWT

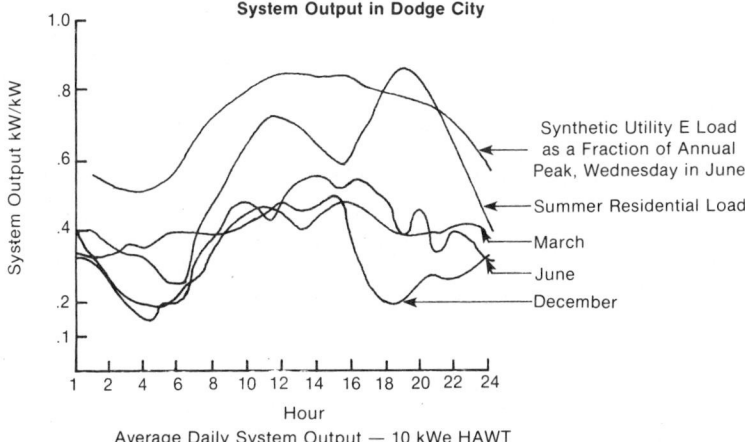

Average Daily System Output — 10 kWe HAWT

FIGURE 7. 10 kWe Horizontal Axis Wind Turbine for Residential Application.

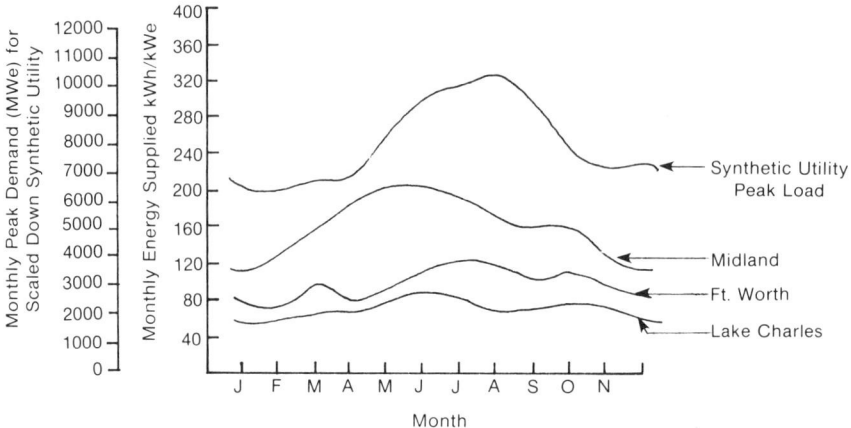

Monthly Energy Supplied by a 100 kWp Line Focus Photovoltaic System

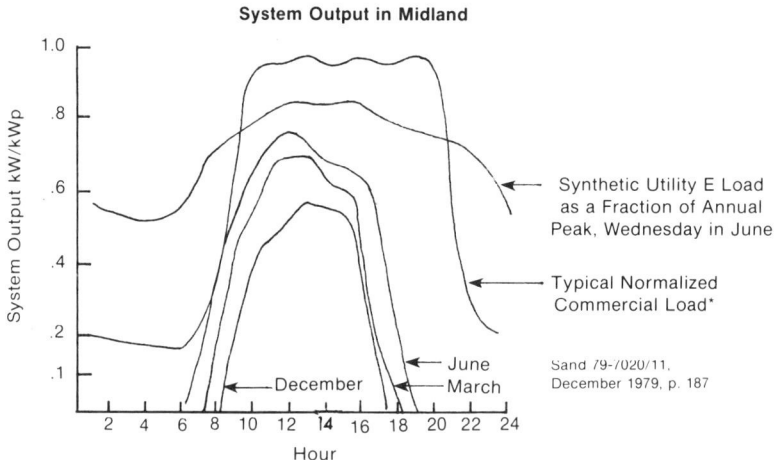

Average Daily System Output — 100 kWp Line Focus Photovoltaic System

FIGURE 8. Line Focus 100 kWp Photovoltaic System for Commercial Application.

a commercial user in this region as well as the load profile for a synthetic utility in the south central region. These data are fed into the first three steps of the economic analysis (see figure 2) to determine:

1. the hourly output of the solar electric system
2. the hourly excess solar energy
3. the hourly makeup of energy from the utility

THE ECONOMICS OF DISPERSED TECHNOLOGIES

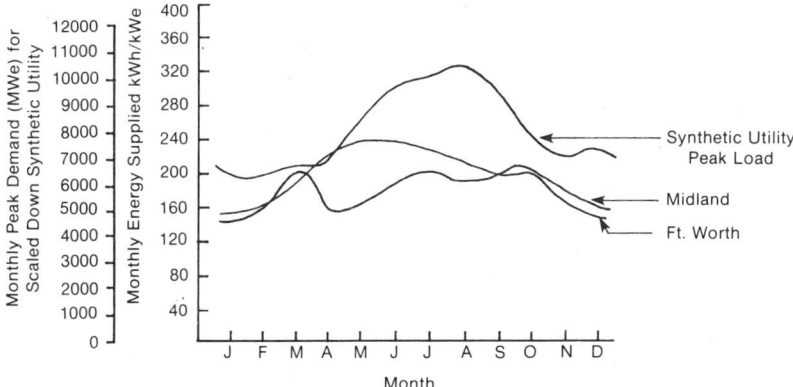

Monthly Energy Supplied by a 210 kWe Dish Organic Rankine Solar Thermal System

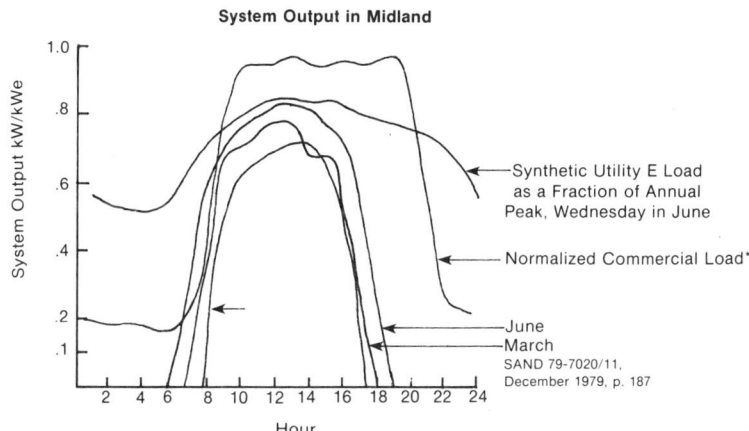

FIGURE 9. 220 kWe Solar Thermal Dish Organic Rankine System, Midland, for Commercial Application.

The information in these figures shows that the solar electric technologies utilizing insolation are better matched with energy demand (by both the user and the utility) than the wind energy conversion systems. This should result in a greater displacement of higher-cost fuels (used at peak) by these systems than by wind systems.

On the other hand, wind machines operate with a higher capacity factor at the prime resource sites within this region, that is, they produce more energy. Tables 7 and 8 summarize the energy produced and the excess energy for sell-

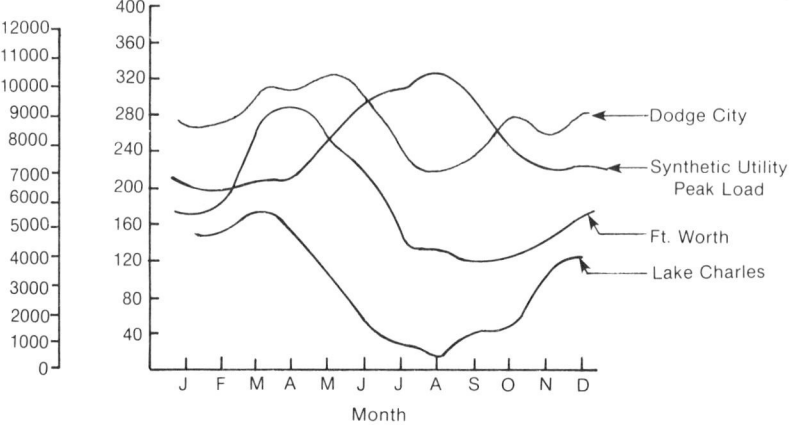

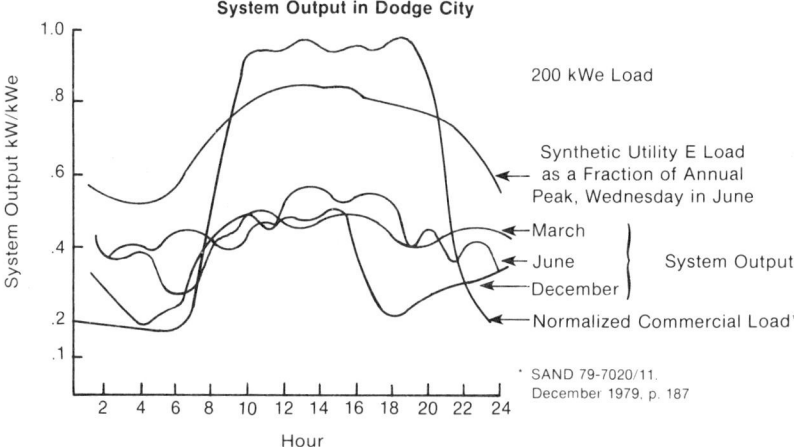

FIGURE 10. 300 kWe Intermediate Horizontal Axis Wind Turbine for Commercial Application.

back to the utility for various regions of the country. The excess energy can range from 0 percent to 60 percent of the energy produced by the given technology. In these tables a convenient measure of the performance of each technology is: kW(e) hour output divided by kW(e) rating of the device. Note that the performance of wind systems is approximately twice as good as PV and solar thermal in most regions. Solar thermal is equal to or better than PV.

TABLE 7. Performance of Dispersed Solar Electric in the South Central Region

Region	Technology	Application	Size	C.F.[a]	Total Energy Delivered		% to Load	% to Utility	Remarks (16. MWh Load)
					kW(e)h	kW(e)h/kW(e)			
South Central	PV	residential	10 KWp	.21	18,400	1840	40	60	(1)
				.17	1,500	1500	42	58	(1)
				.15	13,100	1310	49	51	(3)
South Central	PV	commercial	100 KWp	.19	169,000	1690	100	0	(1)
				.13	112,000	1120	100	0	(2)
				.09	79,000	790	100	0	(3)
South Central	Solar Thermal Dish	commercial	210 KWe	.25	456,750	2175	80	20	(1)
South Central	Wind	residential	10 KWe	.34	30,300	3030	27	73	(4)
				.22	19,200	1920	32	68	(2)
				.11	9,200	920	33	67	(3)
South Central	Wind	commercial	300 KWe	.38	993,000	3310	75	25	(4)
				.25	666,000	2220	87	13	(2)
				.13	339,000	1130	98	2	(3)

a. C.F. = Capacity Factor
(1) Midland, TX; (2) Ft. Worth, TX; (3) Lake Charles, LA; (4) Dodge City, KS.

TABLE 8. Performance of Dispersed Solar Electric in the North Central Region

Region	Technology	Application	Size	C.F.[a]	Total Energy Delivered		% to Load	% to Utility	Remarks (16. MWh Load)
					kW(e)h	kW(e)h/kW(e)			
North Central	PV	Residential	3 kWp	.174	4,124	1,374	54	46	Madison, WI 3.76 kWm/m² day
		Small Commercial	30 kWp	.174	41,237	1,374	91	9	
		Large Commercial	300 kWp	.174	412,373	1,374	94	6	
		Small Ind.	2.15 MWp	.174	2.955 × 10⁶	1,374	91	9	
		Large Ind.	36 MWp	.174	49.48 × 10⁶	1,374	91	9	
		Agricultural	7 kWp	.174	9,623	1,374	77	23	
	Wind	Residential	4 kWe	.42	13,311	3,327	40	60	Rochester, MN 8 M/S
		Small Commercial	40 kWe	.41	129,257	3,231	73	27	
		Large Commercial	2,200 kWe	.55	1.74 × 10⁶	4,350	66	34	
		Small Industrial	2,500 kWe	.44	8.69 × 10⁶	3,476	75	25	
		Large Industrial	14,2,500 kWe	.44	1.21 × 10⁸	3,457	85	15	
		Agricultural	2,4 kWe	.42	26,623	3,327	69	31	

a. C.F. = Capacity Factor

TABLE 8 (Cont'd)

Region	Technology	Application	Size	C.F.[a]	Total Energy Delivered		% to Load	% to Utility	Remarks (16. MWh Load)
					kW(e)h	kW(e)h/kW(e)			
NW	Wind	Residential	4 kWe @ 26 mph	.32	9,990	2,497	70	30	(5) 12**
				.39	12,400	3,100	60	40	(6) 14.3**
				.51	16,200	4,050	61	39	(7) 16.7**
				.41					(5) 14**
	PV	Residential	3.3 kWp	.18	5,200	1,575	73	27	(1)
			7 kWp	.17	10,500	1,500	46	54	(2) 3.96
			3.3 kWp	.13	3,700	1,121	78	22	(3)
			3.3 kWp	.21	6,000	1,818	73	27	(4)
		Low-Rise Apartment	45 kWp	.18	80,300	1,784	45	55	(1)
				.17	67,800	1,506	53	47	(2)
				.13	50,300	1,117	60	40	(3)
				.21	81,600	1,813	49	51	(4)

(1) Omaha, NE; (2) Great Falls, MT; (3) Seattle, WA; (4) Holmes, HI; (5) Goodland, KS; (6) Middleton Is., AK; (7) Kahuku Upper, HI
a. C.F. = Capacity Factor
**Effective Average Annual Wind Speed, mph.

TABLE 8 (Cont'd)

Region	Technology	Application	Size	C.F.[a]	Total Energy Delivered		% to Load	% to Utility	Remarks (16. MWh Load)
					kW(e)h	kW(e)h/kW(e)			
NW	PV	Shopping Center	300 kWp	.18	469,000	1,563	86	14	(1)
				.17	452,000	1,507	84	16	(2)
				.13	335,000	1,117	57	43	(3)
				.21	543,000	1,810	87	13	(4)
		Farm	39.5 kWp	.18	61,800	1,565	50	50	(1)
			45 kWp		67,800	1,506	49	51	(2)
			39.5 kWp	.13	44,200	1,119	57	43	(3)
			39.5 kWp	.21	71,600	1,813	47	53	(4)
NW	Wind	Low Rise Apartment	20 kWe	.51	79,500	3,976	67	33	(1) 12**
				.55	87,200	4,360	63	37	(2) 14.3**
				.71	112,000	5,600	62	38	(3) 16.7**
				.59					(1) 14**

(1) Omaha, NE; (2) Great Falls, MT; (3) Seattle, WA; (4) Holmes, HI
a. C.F. = Capacity Factor
**Effective Average Annual Wind Speed, mph.

TABLE 8 (Cont'd)

Region	Technology	Application	Size	C.F.[a]	Total Energy Delivered kW(e)h	kW(e)h/kW(e)	% to Load	% to Utility	Remarks (16. MWh Load)
NW	Wind	Shopping Center	200 kWe	.42	$.734 \times 10^6$	3,670	62	38	(1) 12**
			(5) 200 kWe	.48	$.835 \times 10^6$	4,175	54	46	(2) 14.3**
				.64	1.118×10^6	5,590	60	40	(3) 16.7**
				.51					(1) 10**
		Paper Mill	2,500 kWe	.32	7.074×10^6	2,829	100	0	(1) 12*
				.41	8.920×10^6	3,568	100	0	(2) 14.3**
				.50	11.0×10^6	4,400	100	0	(3) 16.7**
		Aluminum Plant	20–2,500 kWe 50,000 kWe	.37	1.61×10^8	3,200	100	0	Seven Hill, OR 15.6**
		Farm	20 kWe	.51	79,500	3,975	56	44	(1) 12*
				.71	112,000	5,600	58	42	(3) 16.7*
				.59					(1) 14*

(1) Goodland, KS; (2) Middleton Is., AK; (3) Kahuku, HI
a. C.F. = Capacity Factor
**Effective Average Annual Wind Speed, mph.

TABLE 8 (Cont'd)

Region	Technology	Application	Size	C.F.[a]	Total Energy Delivered kW(e)h	kW(e)h/kW(e)	% to Load	% to Utility	Remarks (16. MWh Load)
NW	Wind	Fluid Milk Plant	200 kWe, 5 units	.42 .48 .64 .51	3.67×10^6 4.18×10^6 5.59×10^6	3,760 4,180 5,590	24 68 69	26 32 31	(1) 12** (2) 14.3** (3) 16.7** (1) 14**

(1) Goodland, KS; (2) Middleton Is., AK; (3) Kahuku Upper, HI
a. C.F. = Capacity Factor
**Effective Average Annual Wind Speed, mph.

REGIONAL VALUE VERSUS COST

In the following paragraphs and charts the value and cost of various solar electric technologies are provided as a function of time for five regions of the United States—the south central region, Hawaii, the Northwest/Plains region, the Pacific Northwest, and the north central region—to provide a range of utilities, solar resources, and load matching.

The South Central Region
The states in the south central region are Texas, Oklahoma, Kansas, Missouri, Arkansas, and Louisiana. However, the sites chosen for analysis represent the range of resources and fuel types available in the southern latitudes of the United States. The regional characteristics pertinent to the analysis are:

1. Electric utilities peak during the summer.
2. Primary fuel used is natural gas (71 percent in 1977, with 18 percent coal in 1977). Projected usage in the year 2000 is 2 percent gas, 68 percent coal; and 17 percent nuclear.
3. Solar resources cover a wide range that is typical of the United States (4–7 kWh/m^2 for direct insolation or 100–600 W/m^2 mean wind power at 50 m above the ground).

The results of an analysis of value and cost for applications in industrial, commercial, and residential sectors[5] are shown in figures 11 through 19. Figures 11 through 13 compare PV with wind at a common site (Fort Worth, Texas) for three dispersed applications by plotting the ratio of value to cost versus time. It can be seen that PV is better than wind for Fort Worth (it has a higher value/cost ratio) for all three dispersed applications.

Figures 14 through 19 plot the actual value for various sites (and the cost) versus time of three applications of various solar technologies. For industrial applications, the value for a 10 percent internal rate of return is shown for three sites and the sensitivity to a 15 percent internal rate of return is shown for a single site. For the commercial and residential applications using life-cycle costing, the results are given for three sites.

For the south central region the results indicate (1) that photovoltaics should reach economic feasibility in the residential market in 1985; (2) that photovoltaics have broader commercialization potential in the south central region than wind systems, although wind systems now have economic feasibility for *high* wind resource sites, such as Dodge City, Kansas (see figures 14, 16, and 18).

Hawaii
Hawaii is a unique market for dispersed solar energy application because it offers excellent resources for both wind and PV (insolation). In addition, utility

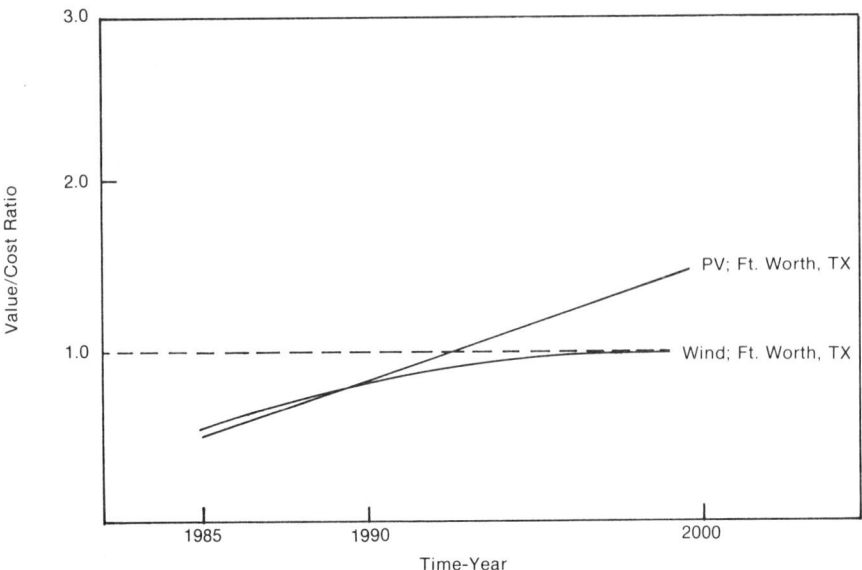

FIGURE 11. Value/Cost Ratios for Industrial Applications in the South Central Region.

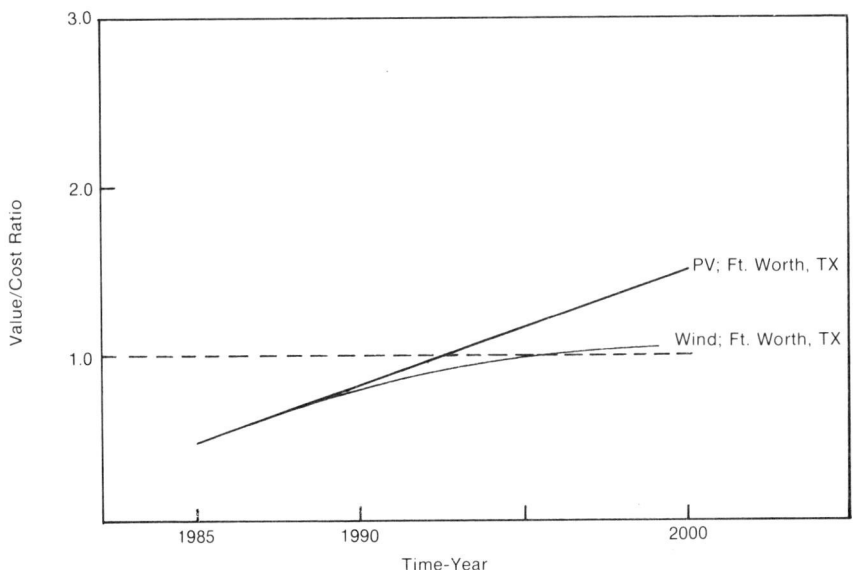

FIGURE 12. Value/Cost Ratios for Commercial Applications in the South Central Region.

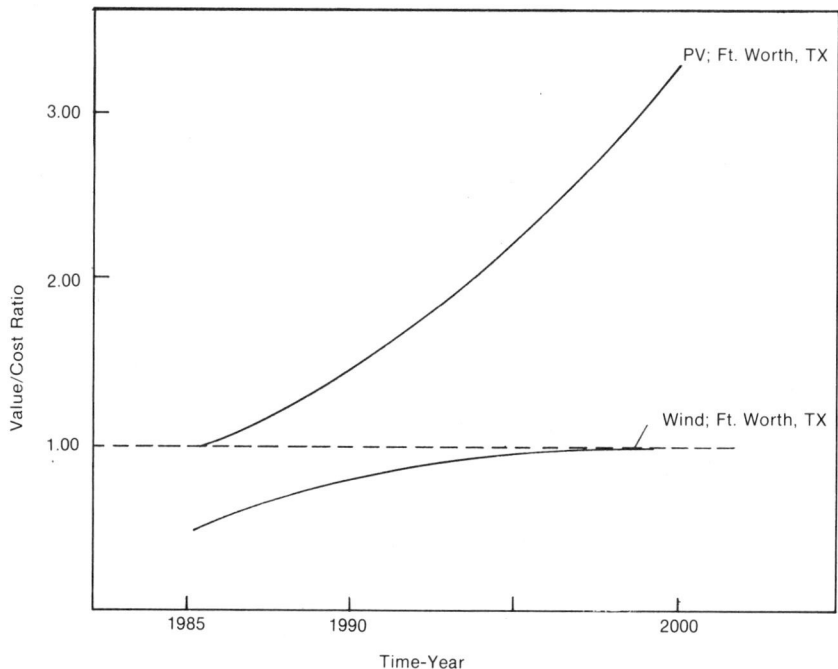

FIGURE 13. Value/Cost Ratios for Residential Applications in the South Central Region.

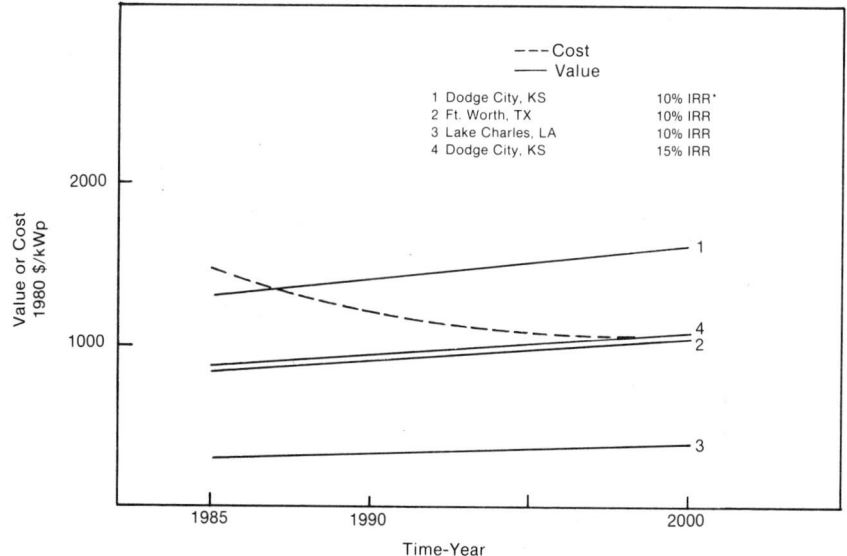

FIGURE 14. Value and Cost for a 300 kWe Horizontal Axis Wind Turbine in an Industrial Application in the South Central Region.

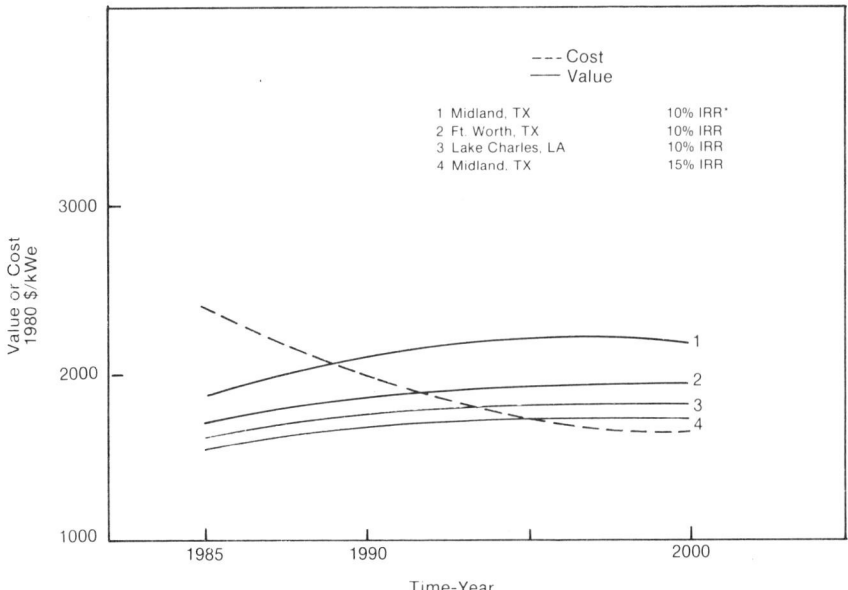

FIGURE 15. Value versus Cost for 100 kWp Flat Panel PV Industrial Application—10% Internal Rate of Return (Midland, TX, also shown with 15% Internal Rate of Return).

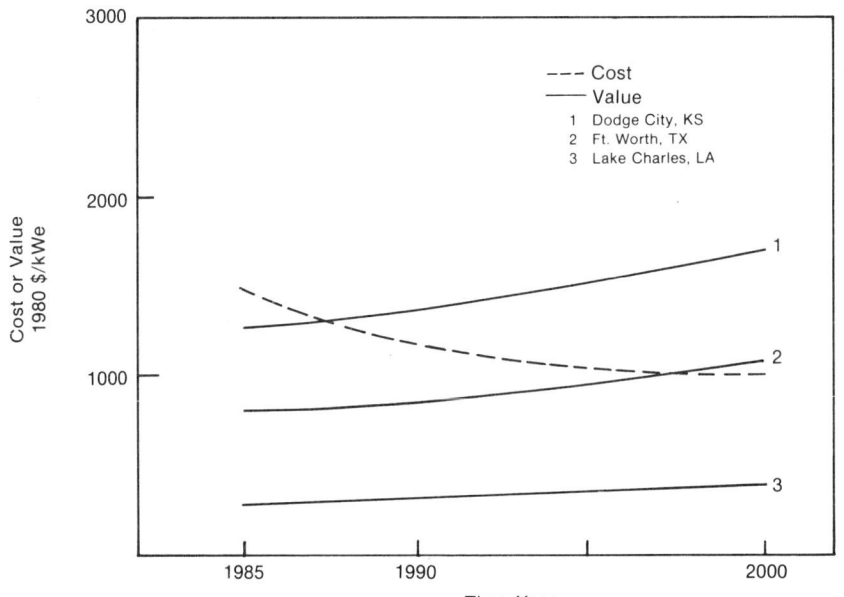

FIGURE 16. Value and Cost for a 300 kWe Horizontal Axis Wind Turbine in Commercial-New Construction Application in the South Central Region.

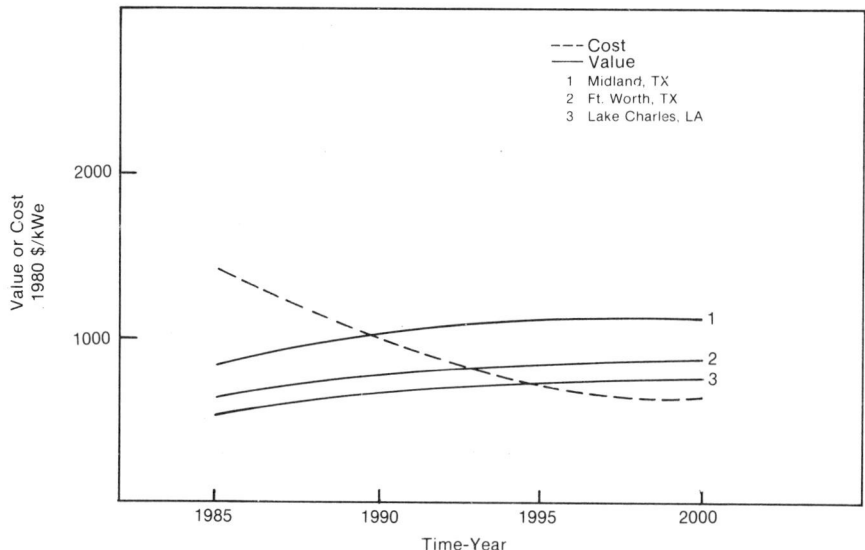

FIGURE 17. Value versus Cost for a 100 kWp Flat Plate Commercial-New Construction Application.

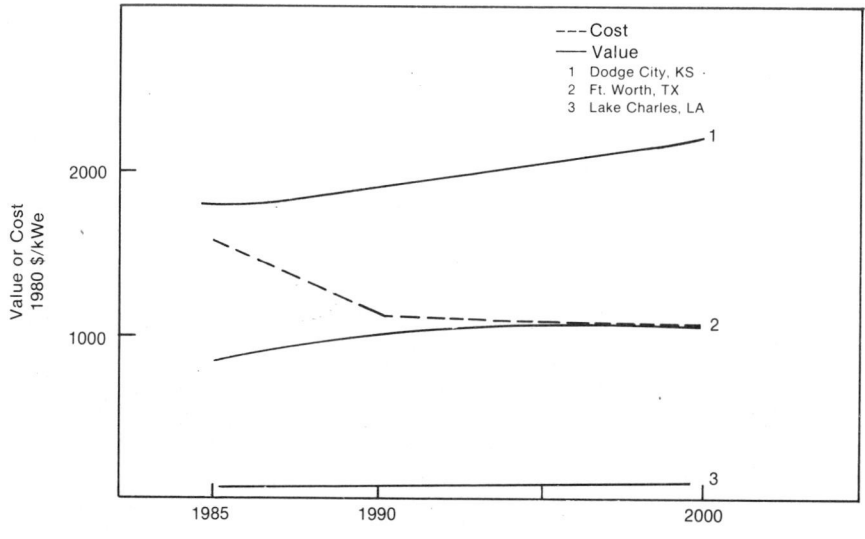

FIGURE 18. Value and Cost for a 10 kWe Horizontal Axis Wind Turbine in Residential Application, South Central Region.

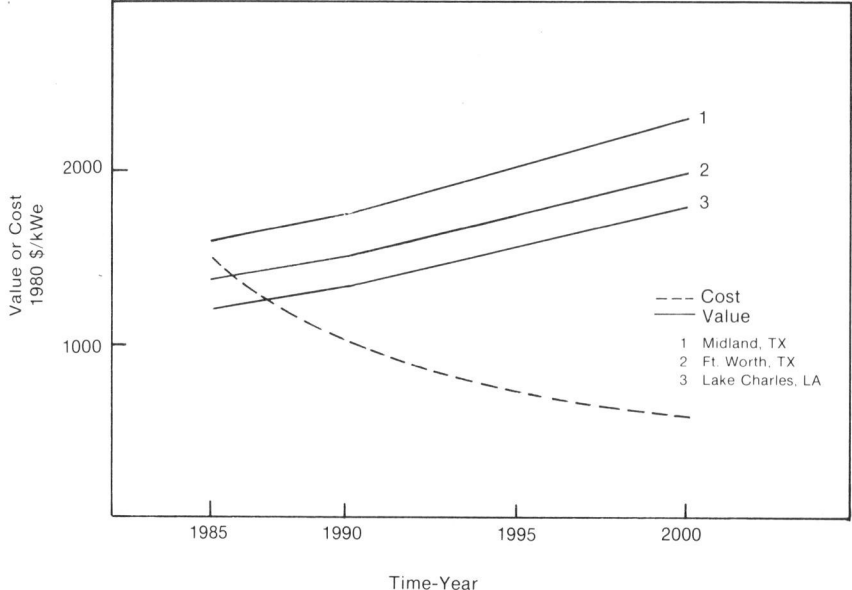

FIGURE 19. Value versus Cost for a 10 kWp Residential Roof-Mounted PV System in the South Central Region.

fuel costs are exceptionally high. The regional characteristics pertinent to the analysis are:

1. Primary fuel used is oil.
2. The state has an active solar energy program.
3. Land cost, availability, and zoning are major concerns.
4. Solar resources are above average (4–7 kWh/m^2 for direct insolation and 400–600 W/m^2 mean wind power at 50 m above the ground).

The results of an analysis of value and cost for applications of various technologies in industrial, commercial, and residential sectors are shown in figures 20 through 22. Figure 20 compares PV with wind for a Kahuku, Hawaii, site and for the three dispersed applications by plotting the ratio of value to cost versus time. Figure 20 indicates that wind systems are better than PV for Hawaii (wind has a higher value/cost ratio) for all three dispersed applications, provided sites can be obtained for wind machines.

Figures 21 and 22 plot the actual value for various sites (and the cost) versus time for the three applications. For all applications, wind systems are already economically feasible while the market for photovoltaics begins to emerge in 1990.

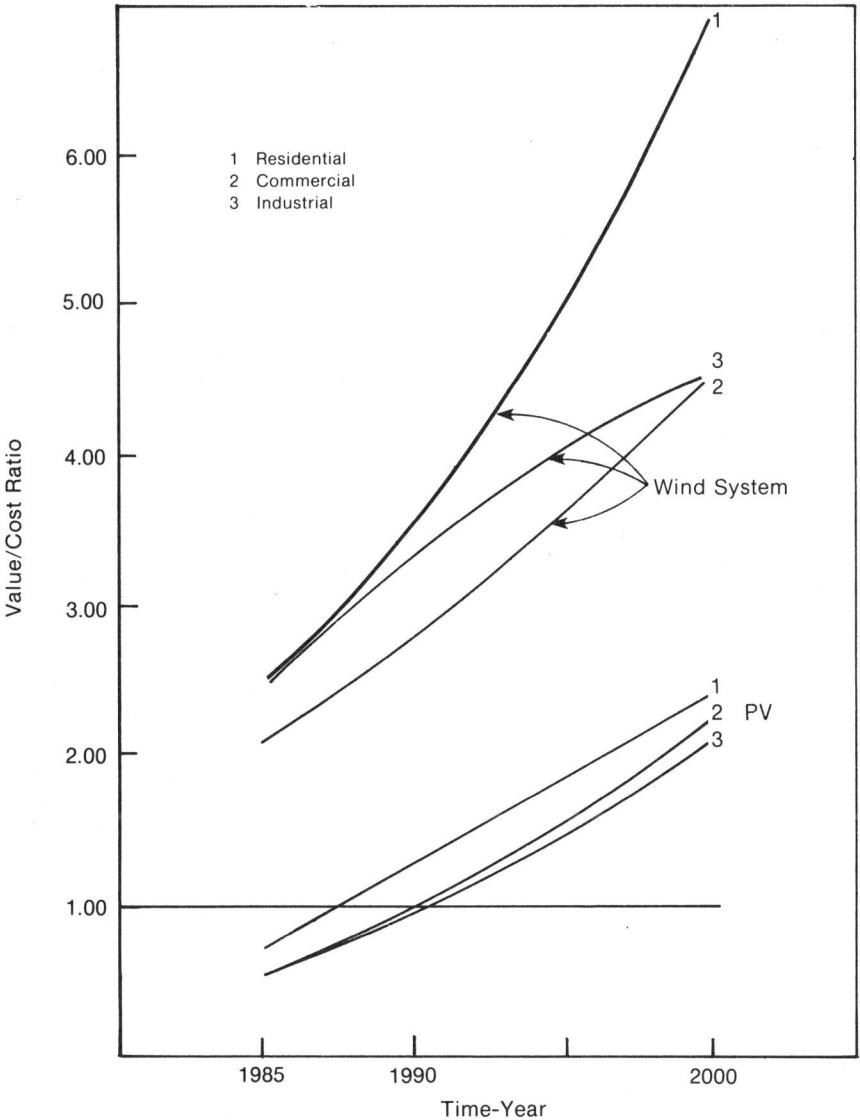

FIGURE 20. Value/Cost Ratios for Wind and PV Dispersed Applications in Hawaii.

For Hawaii the results indicate (1) that dispersed photovoltaics should achieve economic feasibility in 1990; (2) that wind systems have already achieved economic feasibility provided the wind sites are available.

The Northwest
The states forming the Northwest are North Dakota, South Dakota, Nebraska, Wyoming, Montana, Idaho, Oregon, and Washington. The region can be divided

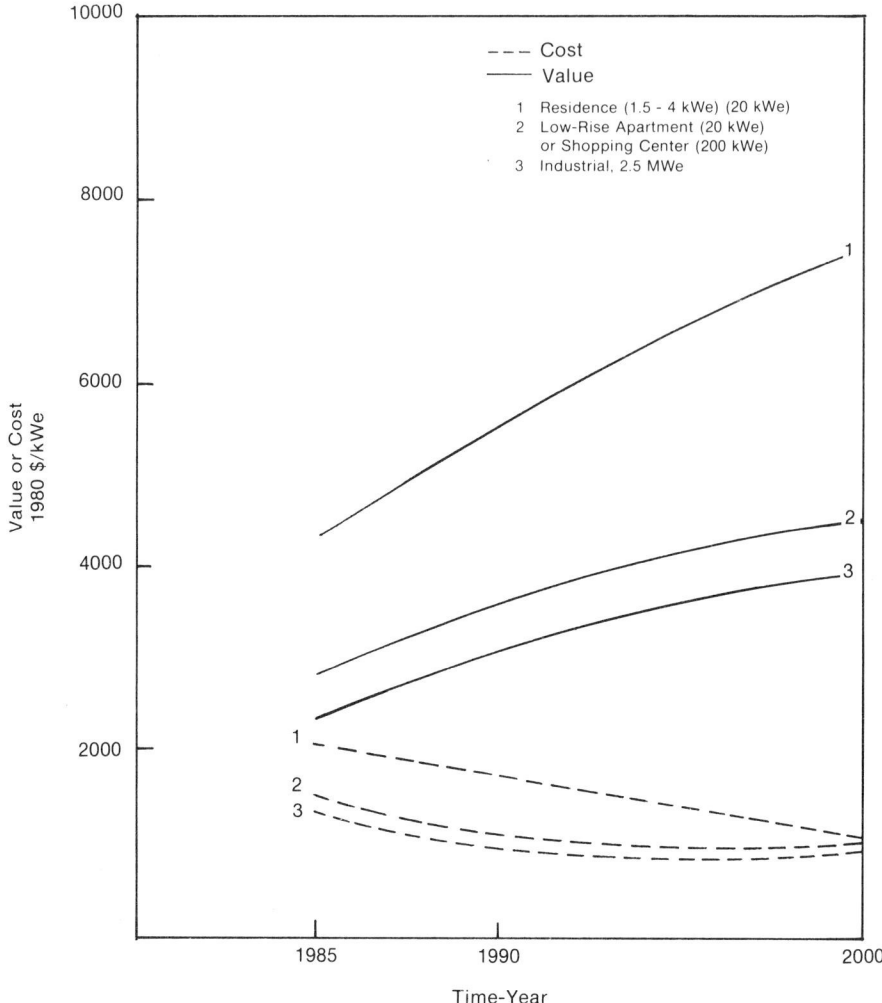

FIGURE 21. Value and Cost for Dispersed Horizontal Axis Wind Turbine in Hawaii.

into two subregions for analysis according to the range of resources and fuel types available. The plains states of North Dakota, South Dakota, and Nebraska form a subregion with the following characteristics:

1. Electric utilities are generally public or municipal with low growth rates.
2. Primary fuel used is coal.
3. Solar resources are average for both wind and insolation (5–6 kWh/m² for direct insolation or 300–350 W/m² mean wind power at 50 m above the ground).

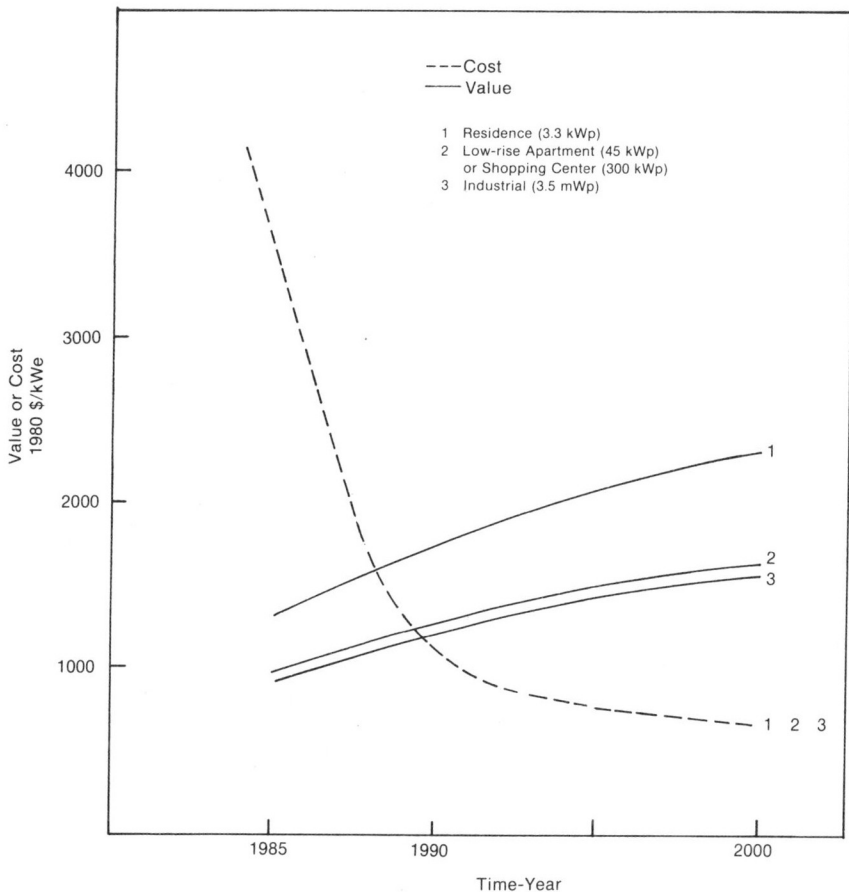

FIGURE 22. Value and Cost for Dispersed Horizontal Axis Wind Turbine in Hawaii.

The states in the second subregion (Pacific Northwest) are Wyoming, Montana, Idaho, Oregon, and Washington. Their subregional characteristics are:

1. Electric utilities are dominated by power pools that provide low-cost electricity.
2. Primary resource used is large-scale hydro generation with a projected shift to coal in the future.
3. Solar resources cover a wide range (2–7 kWh/m^2 for direct insolation or 200–500 W/m^2 mean wind power at 50 m above the ground).
4. Larger population centers are in the Northwest.
5. Positive attitude exists toward dispersed applications.

The results of an analysis of value and cost for applications in industrial, commercial, and residential sectors are shown in figures 23 through 26. Figures

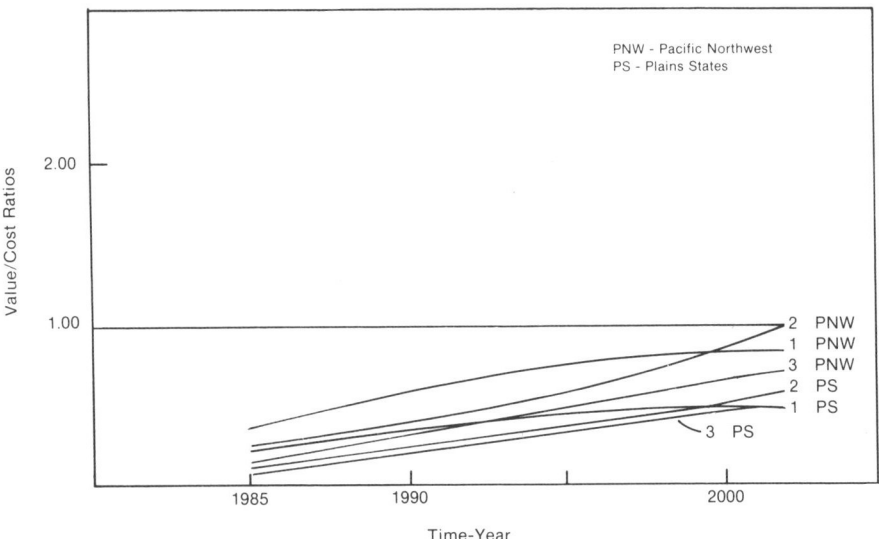

FIGURE 23. Value/Cost Ratio for PV Dispersed Applications in the Northwest Region.

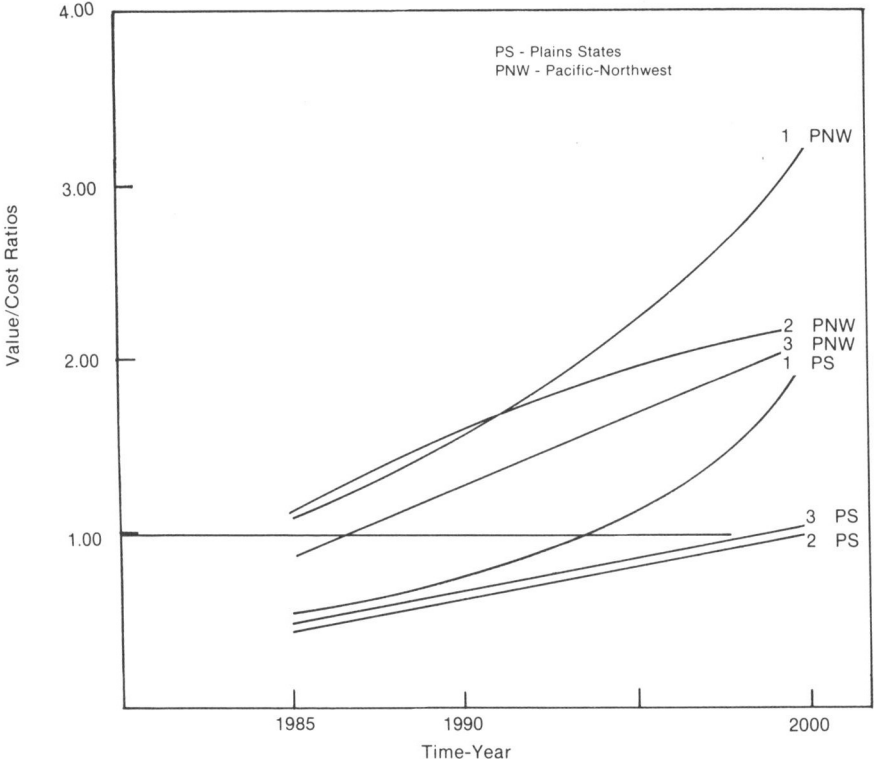

FIGURE 24. Value/Cost Ratio for Dispersed Wind Systems in the Northwest Region.

THE ECONOMICS OF DISPERSED TECHNOLOGIES

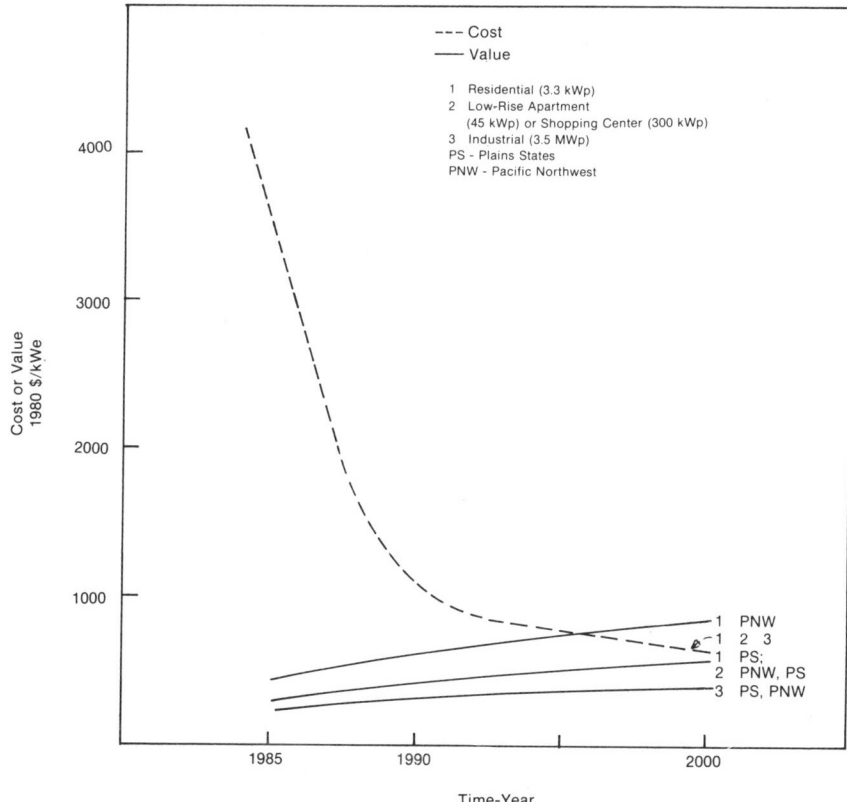

FIGURE 25. Value and Cost of Dispersed Photovoltaics Systems in the Northwest Region.

23 and 24 show PV and wind at common sites for three dispersed applications by plotting the ratio of value to cost versus time. Figure 24 shows that wind systems have already achieved economic feasibility (a value/cost ratio greater than one) for all three dispersed applications in the Pacific Northwest subregion, while figure 23 indicates that PV does not appear to achieve economic feasibility even by the twenty-first century. Figures 25 and 26 plot the actual value for the two subregions (and the cost) versus time for the three applications.

For the Northwest the results indicate (1) that wind systems have reached economic feasibility while PV systems have low market potential; (2) that mediocre solar resources and the ability to use coal and cheap hydro limit the application of dispersed solar electric systems.

The North Central Region

The states of the north central region are Illinois, Indiana, Iowa, Kentucky, Michigan, Minnesota, Ohio, Pennsylvania, West Virginia, and Wisconsin. Unlike

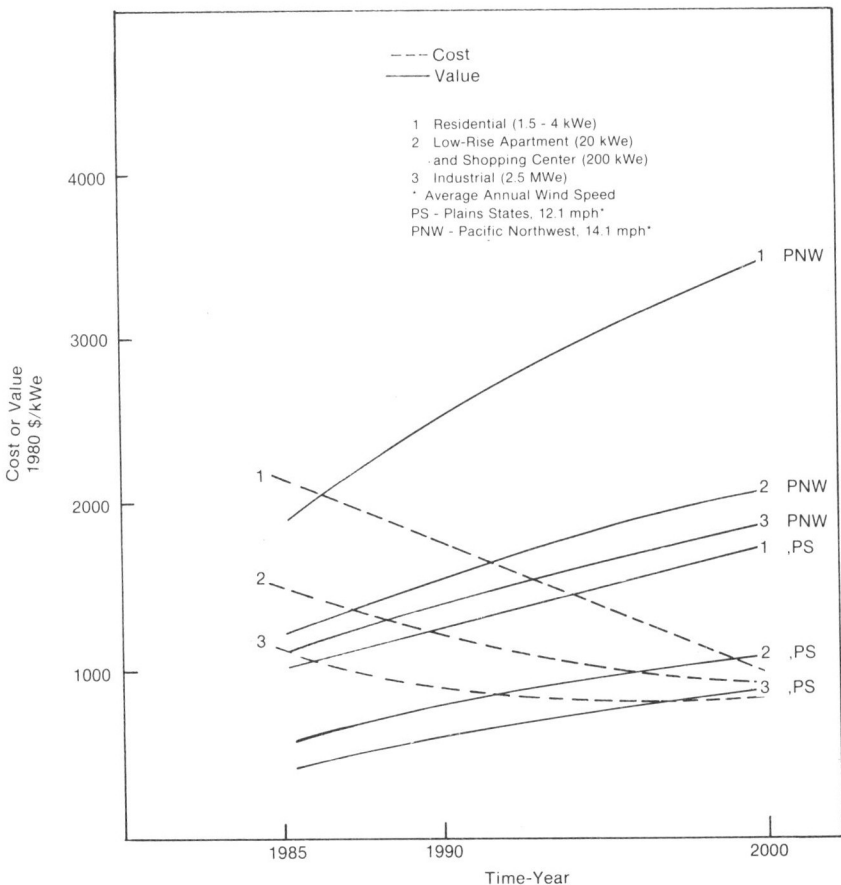

*Average Annual Wind Speed

FIGURE 26. Value and Cost of Dispersed Wind Systems in the Northwest Region.

the Northwest, however, this area is fairly homogeneous in its range of resources and fuel types available. Its pertinent regional characteristics are:

1. Electric utilities peak during the summer.
2. Primary fuel used is coal (95 percent in 1977); only Michigan uses coal and oil (18 percent).
3. Solar resources are homogeneous and only fair (3–4 kWh/m²/day for direct insolation, and 200–300 W/m² mean wind power at 50 m above the ground).

THE ECONOMICS OF DISPERSED TECHNOLOGIES

The results of an analysis of value and cost for applications in industrial, commercial, and residential sectors are shown in figures 27 through 30. Figures 27 and 28 show PV and wind systems for three dispersed applications by plotting the ratio of value to cost versus time. Neither PV nor wind systems are economically feasible in most of the region (both have a value/cost ratio less than 1.00 for all three dispersed applications). However, in Michigan, which uses more oil, both technologies have achieved economic feasibility. Figures 29 and 30 plot the actual value for two principal utilities and the cost versus time for the three applications.

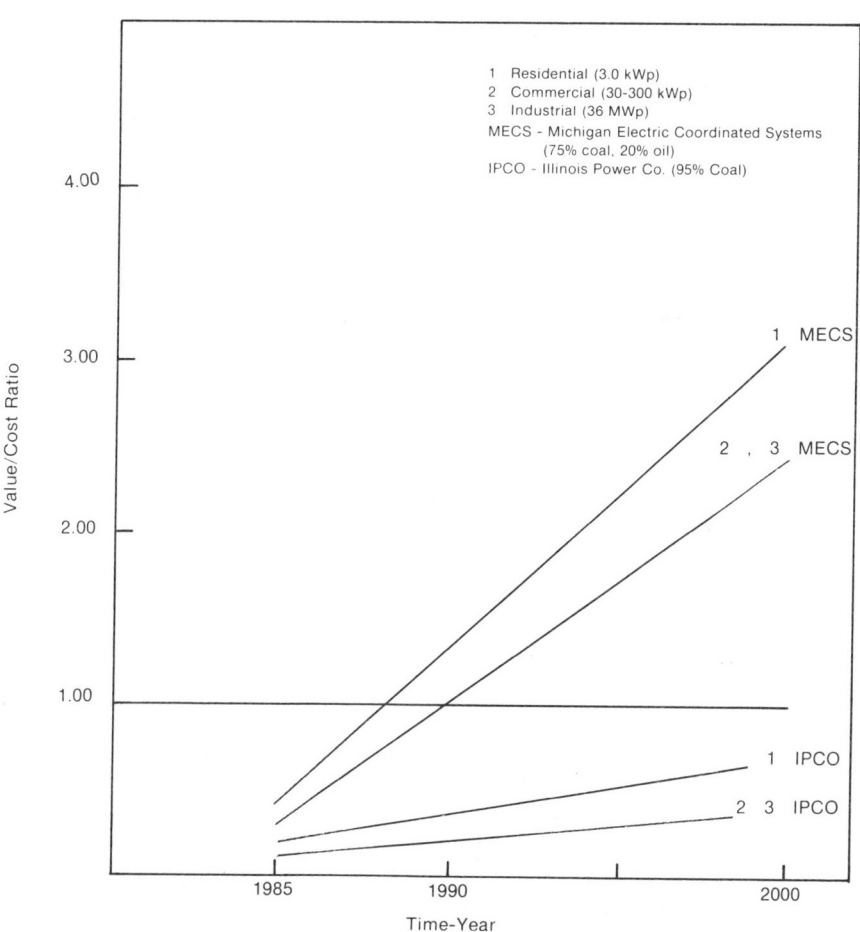

FIGURE 27. Value/Cost Ratio for Dispersed Photovoltaics in the North Central Region.

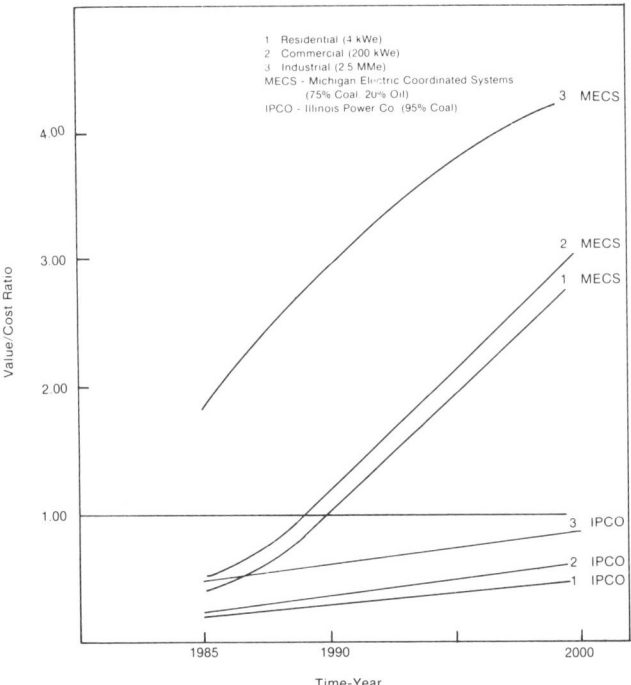

FIGURE 28. Value/Cost Ratio for Dispersed Wind Systems in the North Central Region.

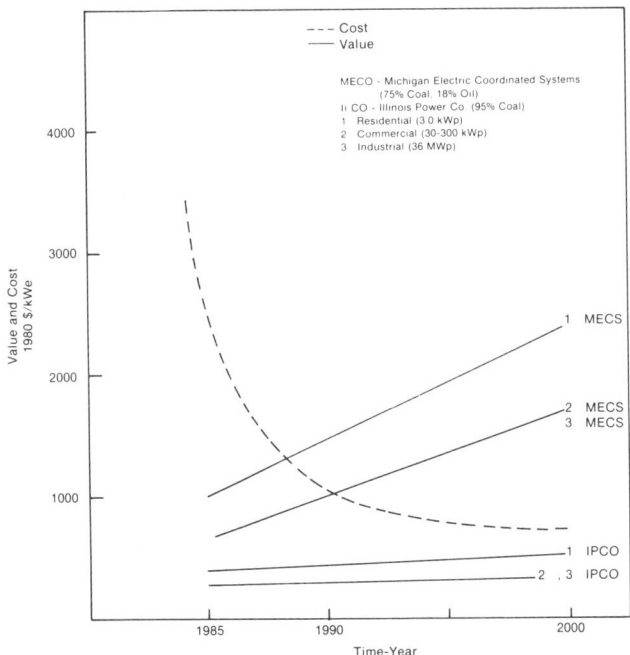

FIGURE 29. Value and Cost of Dispersed Photovoltaics Systems in the North Central Region.

THE ECONOMICS OF DISPERSED TECHNOLOGIES

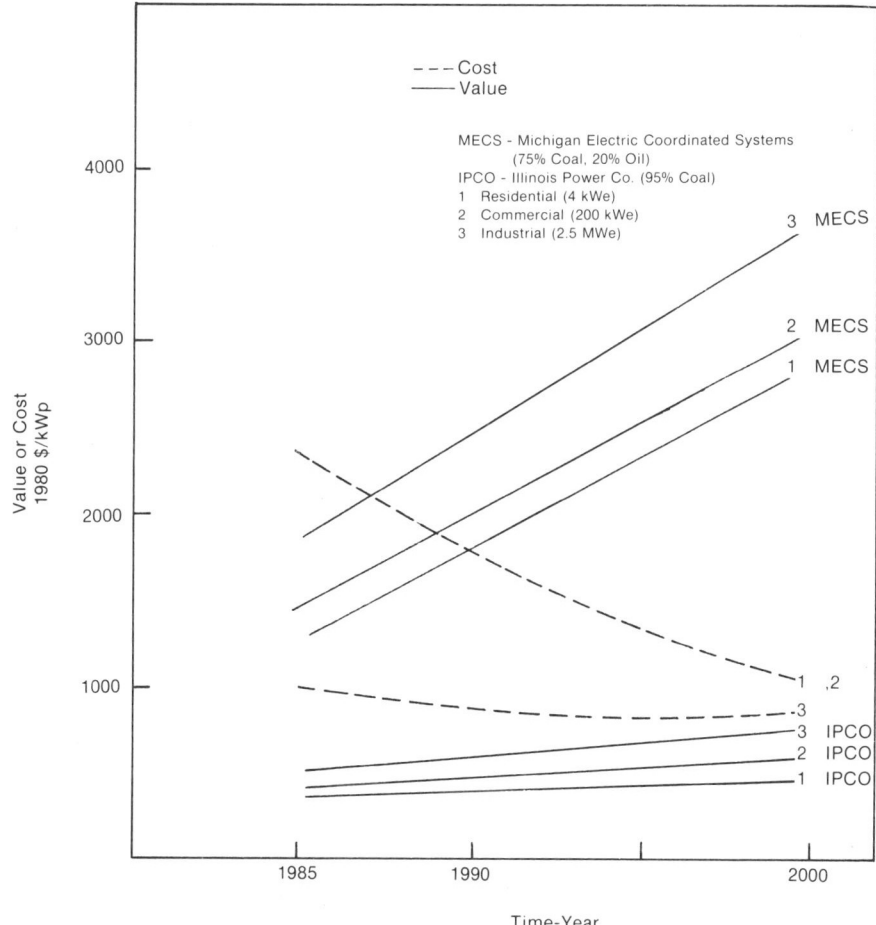

FIGURE 30. Value and Cost of Dispersed Wind Systems in the North Central Region.

For the north central region the results indicate (1) that neither photovoltaics nor wind systems have reached economic feasibility in most of the region because of the dominance of coal usage and mediocre solar resources; (2) that Michigan offers the greatest potential for photovoltaics and wind applications.

NOTES

1. The rate that the utility will pay for excess electricity (referred to as sellback rate) is determined analytically to be the fuel saved by the utility. In reality, there are cases where utilities are paying the full user rate for the excess electricity.
2. A simplified analysis might use annual average insolation data instead. However, an assessment of the backup and sellback portion requires an hourly computation.
3. For a simplified calculation (without the utility probabilistic model) we can use the regional utility rates for the cost of backup energy and 75 percent of this rate for sellback.
4. This cost has removed the estimated O & M cost of the solar technology.
5. Analysis by sector is important because rates vary by sector.

JAMES GUSTAVE KAHN

CHAPTER 8 | *The Advantages of Integrating Decentralized Renewable Electrical Technologies: A Connecticut Case Study*

In 1981 James Kahn completed his thesis at the College of Science in Society at Wesleyan University. It was based on his one-and-a-half-year-study of the potential of integrating decentralized technologies into electrical grids. He collected thousands of data from numerous public and private institutions and analyzed them at Wesleyan's Computer Center. This article is a synopsis of his work.

Chapter 4 assessed the capability of alternative electrical generation technologies to permit a decentralized electricity system in the United States. Kahn extends chapters 4 and 5 by addressing the reliability of such systems. He uses extensive data from a statewide Connecticut study to determine whether the sources can be managed in a synergetic manner to increase reliability. His conclusions are that multiple-site systems make better sense in electricity production than single-site systems; that wind systems benefit highly from being dispersed over a wide geographic area; and that a multiple-site sun/wind/hydro hybrid system is the most reliable.

The most common argument against using decentralized renewable electric technologies (DRETs) in electric systems is that their variability makes them unreliable. Utilities often maintain that DRETs cannot be relied upon to produce power when it is needed or even to produce it consistently over short time periods. Therefore, they claim, the value of DRETs to the utility is limited; some technologies may even cause financial losses because of the costs associated with accommodating short-term fluctuations in output.

Another argument, examined here, is that by dispersing DRETs geographically and by using a variety of technologies, variability may be greatly reduced. Many of the problems and costs of DRETs acting alone can thus be lessened.

A CONNECTICUT CASE STUDY

This study examined the feasibility of integrating solar cells, wind machines, and hydroelectric plants. The word "integrated" here has two aspects. One is the location of DRETs at dispersed sites up to several hundred miles apart. The other is simultaneous reliance on more than one type of renewable energy. The goal of the study was to compare integrated and nonintegrated DRETs to determine their relative suitability for use in an electric system. The following issues are important from a utility company's perspective:

Matching electricity supplied by DRETs with electricity demand on an average daily and yearly basis. If, on the average, electricity is produced by renewables when it is most in demand, much of the capacity requirement of the electrical system can be displaced by renewable systems. If the match is poor (if renewable systems produce more energy during low demand times), the utility company will have to maintain most of its conventional generating capacity for peak demand, though it will still save some fuel and operating costs.

Distribution and variability of power from renewables. Even if average supply and demand are well matched, fluctuations in DRET output could create difficulties in continuously meeting demand. At present, only demand fluctuates beyond utility company control; the supply can be quickly adjusted to meet those fluctuations. If renewable electric supply adds to the fluctuations, the conventional plant adjustment capacity may have to be considerably greater, with higher costs. Three characteristics must be measured in order to determine the extent of the distribution and variability of DRETs.

Distribution of power over various time periods. This indicates to the utility operator how consistent power output is over different lengths of time. The more consistent output is, the less conventional power must be kept ready as reserve to counter the renewables' variations.

Predictability of power for various lengths of time. Weather reports are unreliable for forecasting sun and wind conditions. "Predictability" as defined in this study is the accuracy with which power output can be predicted at a given hour by knowing what power output is at one of several previous hours. It is important because utilities need lead time to adjust other capacity to the variations of DRETs. The more predictable DRET power is, the more intermediate and baseload power plants can be used instead of peaking plants, which are rapidly adjustable but expensive.

Persistence of high and low power production and the need for storage. This measure indicates how long (expressed in days) periods of very low or very high output persist. It is important in energy systems (of any size) with a very high DRET contribution. The less persistence there is, the less need there is for storage to help match supply and demand. This is true because when energy production from DRETs oscillates rapidly between highs and lows, only less expensive, short-term storage is necessary. Conversely, when long periods of extreme highs or lows occur, expensive, long-term storage is required.

INTEGRATING DECENTRALIZED TECHNOLOGIES

Numerous studies have examined aspects of the issues outlined above. The distinguishing features of this study are the following:

Many renewable configurations are considered for each statistical measure: single-site wind, sun, hybrid, and sometimes hydroelectric, as well as multiple-site arrangements of the same sources.

Hybrid systems in particular are modeled at multiple sites. Other studies have examined multiple-site single-source systems and single-site multiple-source systems but generally not both aspects at once. This is significant because multiple locations with multiple sources may prove to be the best combination.[1]

The hybrid and multiple-site effects for wind and sun are analyzed for a wide range of different time scales—years, months, days, as well as multiple and single hours. Thus, findings for short and medium time periods can be derived. Most studies deal only with days and longer time periods.[2]

Meteorological data used for analyzing "simultaneous" DRET power production are actually from the same hour, and those used for analyzing "single-site" hybrid systems are from one site. Many other studies have relied on data from sites ten to a hundred miles apart, or on wind and sun data from different years, or both.[3] Conclusions are then based on statistical assumptions unnecessary in this study.

Hydroelectric power is included in the analyses of hybrid systems carried out on a daily and monthly time basis. Most other studies have not included hydroelectric power.

Predictability of power is examined for seven carefully selected pairs of hours. Few studies examine predictability, and of those that do most are not so oriented to utility needs.[4]

METHODOLOGY

Collecting Data

Two basic types of data were required for the study: meteorological and electrical demand.

Meteorological data for Connecticut are maintained by several government and private organizations. Northeast Utilities Company (NU) has an extensive monitoring network for wind and solar data.[5] The network was chosen as the main data source for many reasons: its data are reliable and complete (95% data recovery); data are stored on computer tapes (ease of use); data are recorded every fifteen minutes (necessary for some measures); wind speeds are higher than other inland data; four sites were covered during the test year; and hybrid data are included at three single sites. The data represent 15-minute averages of wind speed and total horizontal insolation for March, June, September, and December of 1976 (though data for all of 1976 and 1977 were acquired). Solar data were gathered at three sites and wind data at those three plus one other.

National Weather Service data for Bridgeport and Hartford were used to supplement NU's records for wind speeds (eight times per day, one-minute averages) and for average daily temperatures.[6]

Connecticut Department of Environmental Protection data at twelve sites were examined but were considered too low in wind speed and possibly unreliable.[7] Because coastal wind speeds acquired from the *Coastal Zone Wind Energy Study* (funded by the Department of Energy) were available (and acquired) only for earlier years, they could not be mixed with test-year data.[8]

For hydroelectric power, United States Geological Survey records for three gauging stations were used. The data described daily mean stream flows for each site.[9]

Electricity demand data were based on a model developed for NU by the New England Energy Pool (NEPOOL).[10] The model contains historically derived demand equations for each of the twelve months, with four-day types per week in each of those months. Residential, commercial, and industrial demand are separately modeled. Appliances, industrial classifications, temperature-sensitive and non-temperature-sensitive demand, and other sectoral breakdowns are also maintained. To re-create demand shapes, the study used two projected temperature simulations for each month: an average day and a cold or hot day, whichever would create higher demand for electricity.

Processing Data to Simulate DRET Configurations

The data were transformed to simulate renewable electricity production systems. Fortran and SPSS computer programs were used on DEC–10 and DEC–20 computers. Energy output was computed for each technology with equations representing the characteristics of the equipment and its interaction with insolation, wind speed, and stream flow.[11]

In order to facilitate comparison and combination of the different renewables, power output was converted into a standardized form: actual output from a conversion facility divided by output if the facility operated at maximum (installed) capacity all the time. This fraction, known as the *capacity factor*, can apply to any length of time. It allows easy control of the ratio of different sources. For example, since the wind power capacity factor is about twice as high as that of solar cells, if the installed capacity is assumed equal the wind system will generate twice as much energy. The energy output of the solar cells can be made equal by setting their installed capacity equal to twice that of the wind machines. This approach is used for the sun/wind hybrids in the study. The capacity factor methodology is straightforward, generally accepted as valid, and appropriate for all but one of the statistical measures in the study.

For measuring the distribution of solar electricity, a capacity factor is misleading because there is no solar power at night: the capacity factor will be too low on the average. An alternate measure was developed: the *possible power capacity factor* (PPCF) is based on maximum solar output at each hour of the day, one day for each month. This maximum takes into account the angle of solar incidence and estimates atmospheric interference. PPCF equals power output divided by the maximum. A measure based on average power output was rejected because of (1) the desire to use a percentage of a maximum in keeping with the definition of a capacity factor and (2) computer programming convenience. PPCF thus measures the possible availability of power instead of the availability of power as related to equipment characteristics.

Analysis of the DRET System
Various configurations of DRET power output were characterized using statistical measures of distribution, predictability, and persistence of power output patterns. Although I have not included the computer programming details, the meaning and significance of each measure are explained in the next section.

RESULTS

The findings of the study can be divided into several categories: average power over a year; average demand and power over a day; distribution of power; predictability of power; and persistence of high and low power production.

Average Power over a Year
The annual curve of power output is formed by averaging DRET power for each month. Either four sample months or all twelve months per year were used for this measure, depending on the renewable technology. Various DRET configuration curves are reported below and shown in figure 1.

Solar cells are assumed to be mounted to yield a yearly average capacity factor of 12.7%. This is slightly lower than the 16% found in other studies;[12] the difference may perhaps be explained by the ratio of direct and indirect sunshine assumed in different models. The June capacity factor is about 16% and the December capacity factor is only 8%. September (10%) and March (11%) have moderate capacity factors. There is little difference among sites in Connecticut.

Wind machines have outputs that follow the pattern of average wind speed: high in winter and low in summer. The yearlong average is 28%, with a December–March peak of 34% and a June–September low of 21%. Other studies have found higher capacity factors, up to 35%.[13] Different sites in this study have capacity factor averages and pattern variations that are distinct but not very large.

Hybrid sun/wind systems have a much more even yearly output pattern. The average capacity factor is 17.4%, with a March high of 20% and a September low of 14%. For electric systems with fairly even demand throughout the year or with equal peaks in summer and winter, integrated systems would better match demand and would therefore defray a larger portion of conventional capacity.

Hydroelectric power, not included in the above hybrid system, averages 28.6%. It peaks sharply in early spring (March, 56%), when the snow is melting. The low point is in September (10%), when the leaves are still on the trees (evaporating water) and groundwater reserves have dwindled over the summer. June has a slightly higher capacity factor, about 12%, and December's is about 25%. If hydroelectric power were combined with sun and wind systems, March

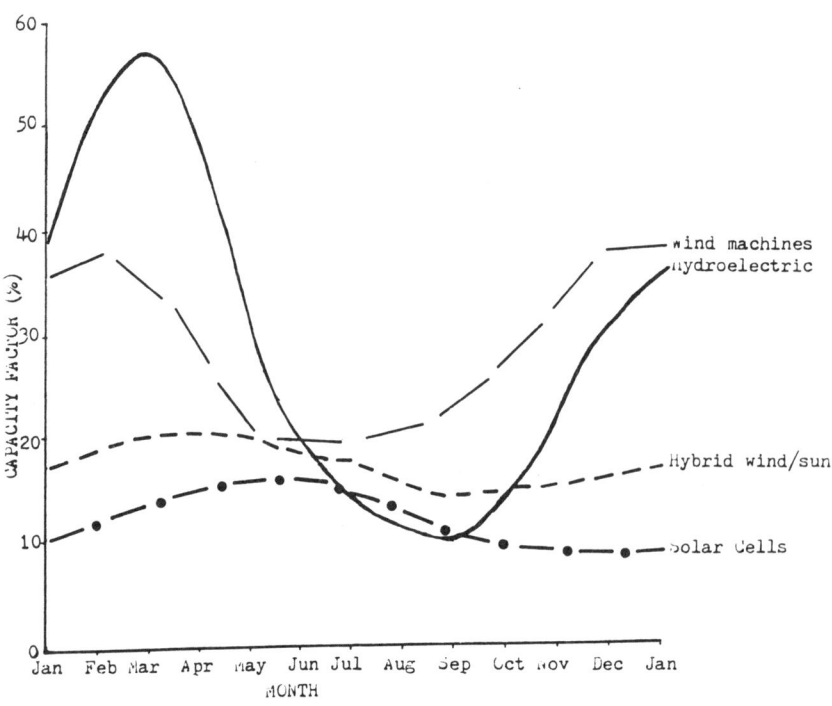

FIGURE 1. Annual Power Curves for Renewable Electric Sources. Month versus Capacity Factor. Note greater evenness of hybrid. Source: James Kahn, "Decentralized Renewable Electricity Generation: Utility System Considerations and Connecticut Reliability Study" (masters thesis, Wesleyan University, 1981).

and December would have slightly higher capacity factors, June and September slightly lower ones. The curve would be almost as smooth over the year.

Average Demand and Power per Day

The average demand per day is based on hourly NEPOOL model equations. The average power per day is based on fifteen-minute intervals averaged over the days of the month; these ninety-six averages create a smooth curve.

Electrical demand generally peaks in the late afternoon (or early evening) and late morning, with evening peaks in December and June (see figure 2). Peaks are caused largely by the residential sector, from heating or air conditioning increases when people return from work, from cooking at specific times, and from water heating (see figure 3). Certain electrical uses that contribute significantly to the peaks may be flexible, most notably clothes drying and hot water use, and could therefore be modified with load management. It is against this average curve that the power production of renewable technologies is compared.

Average power shapes are illustrated in figure 4.

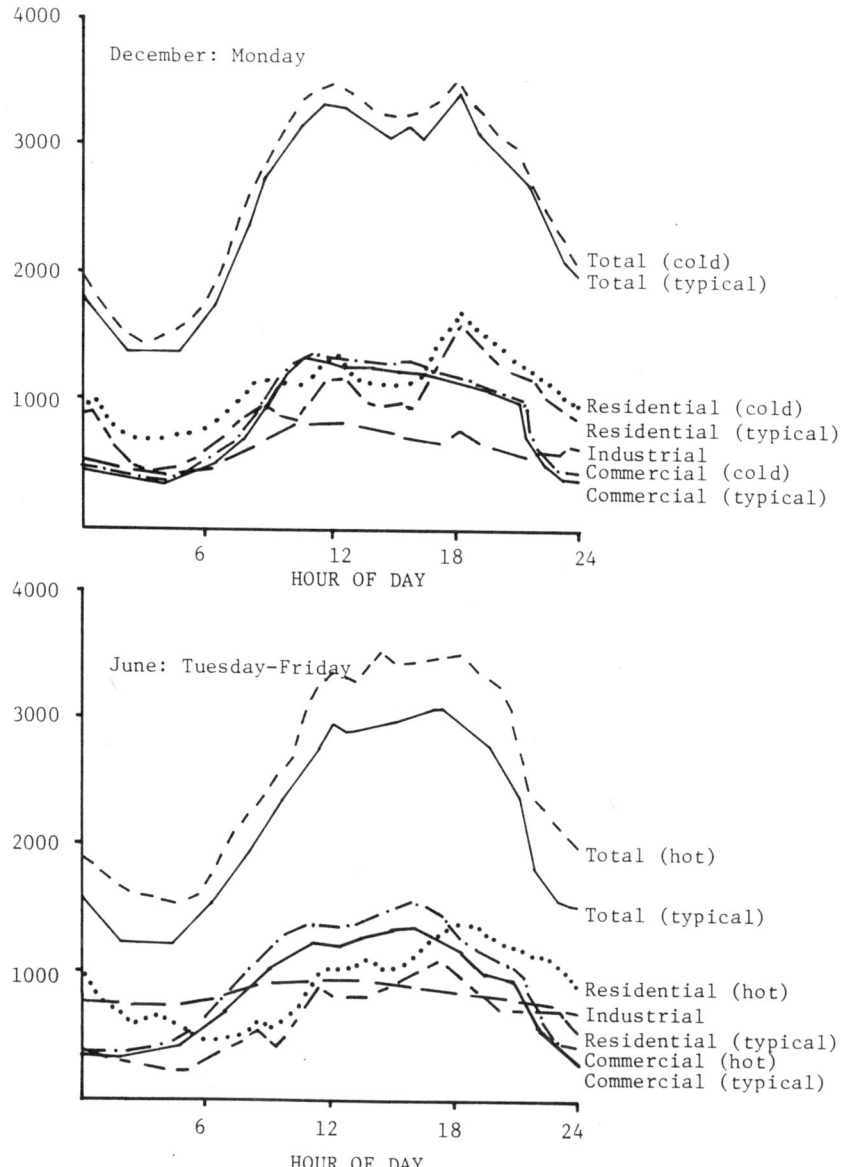

FIGURE 2. Three-Sector Electricity Demand Curves, 1980. Model for Northeast Utilities. Hour of Day versus Megawatts Demand. Source: James Kahn, "Decentralized Renewable Electricity Generation: Utility System Considerations and Connecticut Reliability Study" (masters thesis, Wesleyan University, 1981).

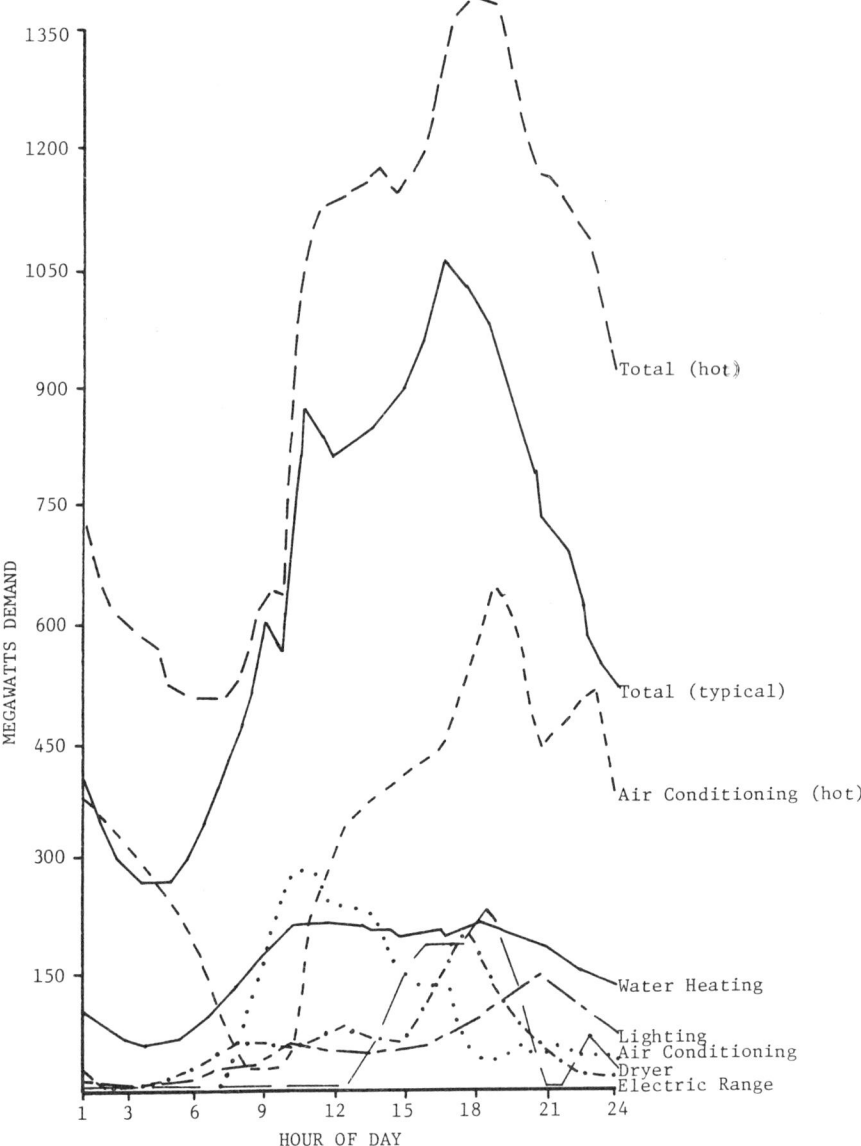

FIGURE 3. Residential Electricity Demand Curves, 1980. Model for Northeast Utilities. Hour of Day versus Megawatts Demand. Note Hour 18 peak and role of air conditioning and range in that peak. Source: James Kahn, "Decentralized Renewable Electricity Generation: Utility System Considerations and Connecticut Reliability Study" (masters thesis, Wesleyan University, 1981).

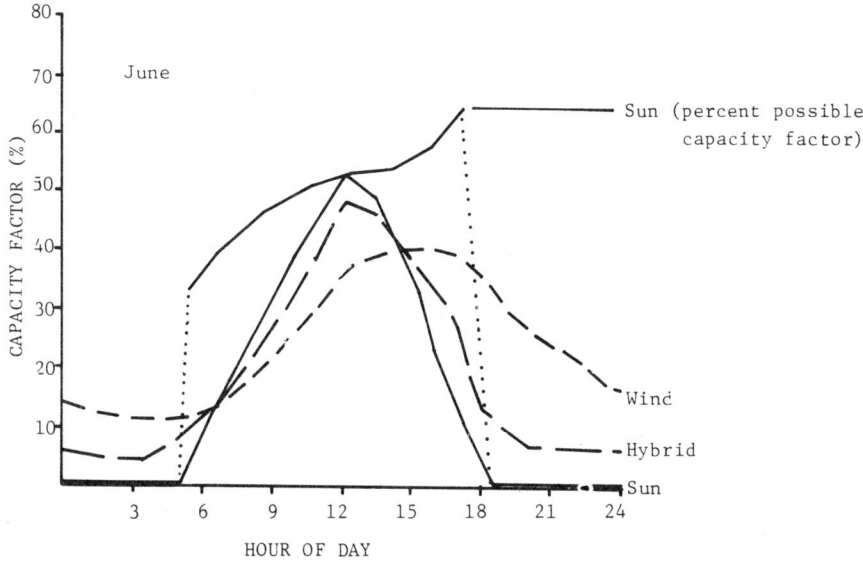
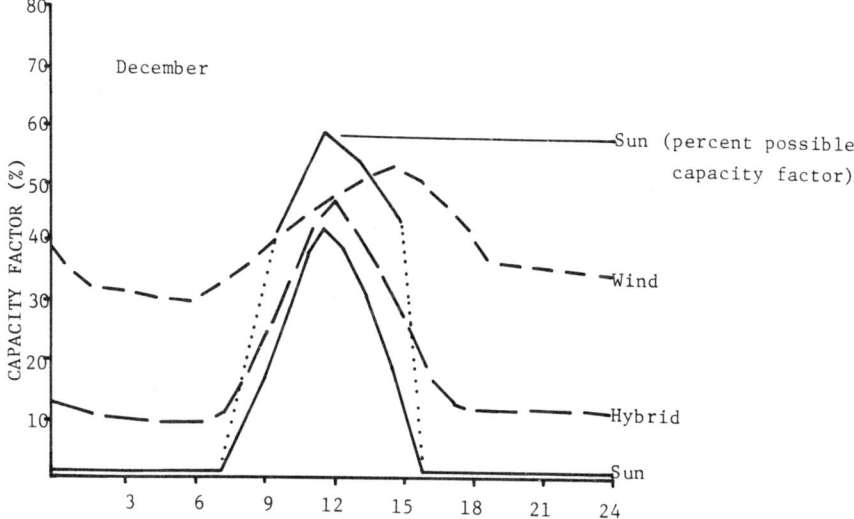

FIGURE 4. Diurnal Power Curves for Renewable Electricity Sources (June and December). Hour of Day versus Capacity Factor. Source: James Kahn, "Decentralized Renewable Electricity Generation: Utility System Considerations and Connecticut Reliability Study" (masters thesis, Wesleyan University, 1981).

Solar cells peak at about noon in all four months, reaching a capacity factor of 40% to 53%, depending on the month. No power is produced from about thirty minutes before sunset to thirty minutes after sunrise. Solar power generally peaks well before the afternoon peak in demand and at the middle of the earlier peak, if any.

Possible power capacity factor (PPCF) reaches 50% to 60% during its peak (also about noon) but remains more constant than capacity factor for most of the sunny portion of the day. PPCF trails off rapidly right before sunset, and after sunrise it rises rapidly.

Wind machines peak in output during the early to mid-afternoon, depending on the month, reaching a high of 40% to 54%. The low occurs between 5:00 and 7:00 A.M., ranging from 12% to 40% in different months. Generally, wind power peaks before the afternoon peak in demand (by as much as several hours or as little as two hours) and after the morning peak (if any).

Hybrid sun/wind systems peak in the early afternoon and reach the low point in the early morning hours, just before sunrise. The peak varies from 45% to 50%, and the low from 4% to 10%.

In summary, power output from sun and wind systems is fairly well matched with electrical demand. The only problem is that the early peak in demand sometimes occurs before renewables are producing at full force. This could be corrected with time-of-day rates, controlled appliance use, or other load management techniques. Once demand occurs after supply, short-term storage could bridge the difference. Clothes drying and hot water use (or water heating in preparation for use) should be the emphasis of the effort to shift demand.

Distribution of Power
Two measures of the distribution of power are reported: percentage of times production is near the mean and percentage of times production is near zero. Near the mean is defined as being within .4 (40%) of the mean from the mean (or about .2 of the maximum from the mean), and being near zero is defined as being within .1 of the maximum from zero. The following guidelines help interpretation:

> less than 20% of cases within .4 of the mean indicates a very non-mean oriented distribution;
> about 40% within .4 of the mean indicates approximately even distribution;
> close to 80% within .4 of the mean indicates a tight mean-centered distribution.

Tighter distribution (closer to the mean) is better for electrical system operation because it indicates greater reliability.

Figure 5 summarizes the findings for distribution of power for different power-averaging time periods.

Solar cells have the most desirable and reliable (mean-oriented) distribution. Over the four months, on the average, 44% of one-hour outputs are within .4 of the mean. Larger averaging times increase the portion near the mean (to 56%). Similarly, the portion within .1 of zero decreases (from 13% to 8%); this

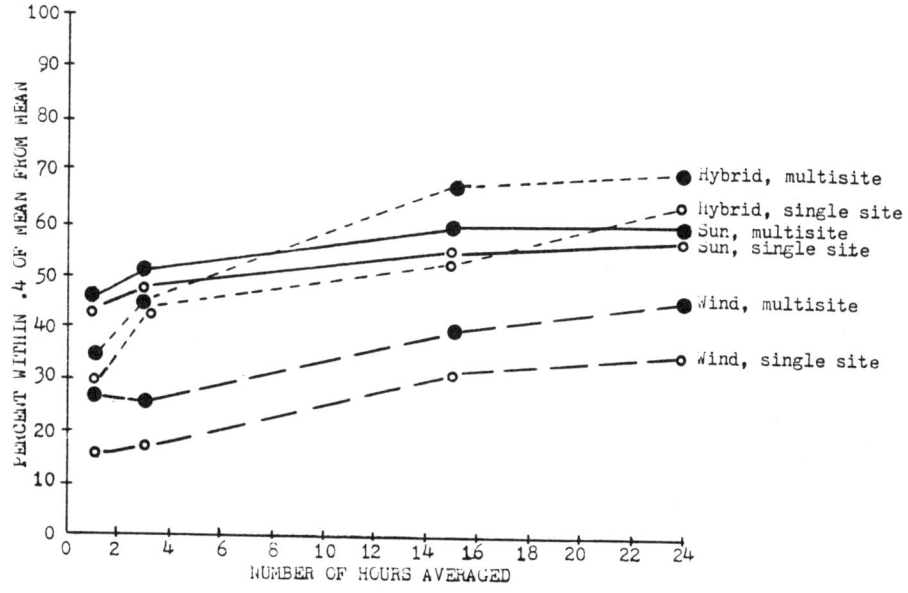

# of Hours Averaged	Single Site		Multiple Site	
	% Within .4 of Mean	% Within .1 of Zero	% Within .4 of Mean	% Within .1 of Zero
SOLAR CELLS				
1	44	13	46	9
3	47	11	51	8
15	54	9	59	7
24	56	8	59	7
WIND MACHINES				
1	15	41	26	25
3	17	40	26	26
15	31	21	39	11
24	34	19	45	11
HYBRID WIND/SUN				
1	29	23	35	12
3	44	11	45	7
15	52	6	67	4
24	63	6	70	4

FIGURE 5. Distribution of Renewable Electrical Production (graph and table). Number of Hours Averaged versus Percent within .4 of Mean. Higher Percent means more concentrated distribution. Hybrid multisite is best for longer averages, sun multisite is best for shorter averages. Source: James Kahn, "Decentralized Renewable Electricity Generation: Utility System Considerations and Connecticut Reliability Study" (masters thesis, Wesleyan University, 1981).

measure is less reliable for solar cells because of inaccuracies in PPCF near sunrise and sunset. Multiple sites have a slightly better distribution, ranging from 46% to 59% within .4 of the mean. The summer months demonstrate better distribution than the winter months.

Wind machines exhibit a wider distribution (are less reliable) but also show greater improvements as a result of longer time averaging or combining multiple sites. The single site has a very spread out (poor) power distribution (15% of cases within .4 of the mean and 41% of cases within .1 of zero for one-hour distribution). By 24 hours the portion within .4 of the mean is increased to 34%, with only 19% within .1 of zero. For the multiple-site array a single hour has 26% of cases within .4 of the mean and 25% of cases within .1 of zero. These measures reach 45% and 11%, respectively, by 24 hours.

Hybrid sun/wind systems exhibit distribution characteristics between those of wind and sun systems. For a single site the one-hour distribution is better than that of the wind machines but worse than that of solar cells (29% within .4 of the mean, 23% within .1 of zero). By 24 hours the single site has better distribution than either single-source system (63% within .4 of the mean, 6% within .1 of zero). The multiple-site hybrid system is also between the two single-source multiple-site systems at one hour and better than either at 24 hours. It reaches a yearlong average of 70% of cases within .4 of the mean and only 4% within .1 of zero. Therefore, hybrid multiple-site systems are better than any single-source or single-site configuration for electric system application.

Predictability of Power
This measure of the change of power over a period of time uses the Standard Error of Estimate (SEE). The SEE indicates the size of the power range that can be expected 68% of the time: 32% of the time estimates will be worse than the SEE, 68% of the time they will be better. SEE is expressed here as a fraction of the average power at the "target" hour (the hour for which power is being predicted). A lower SEE is better, indicating that the range of error is smaller. An SEE of .2 is quite good, .5 mediocre, and anything higher indicates that the prediction is not very useful. An SEE of .2 means that 68% of the predictions are within 20% of the target hour's mean from the real target-hour power output.

Figure 6 summarizes the findings for predictability of power. Two target hours are reported, as suggested by scientists at Northeast Utilities: 1:00 and 4:00 P.M. For each target hour, several lengths of prediction are given, ranging from 1 to 24 hours.

Solar cells are the most consistent in output from one hour to the target hour. The SEE is .3 for a single hour, indicating that two-thirds of the time a prediction made one hour earlier will be within 30% of the average for the target hour. For several-hour predictions the SEE increases to .43, reaching .5 for 20-hour predictions. Twenty-four-hour predictions are slightly better than 20-hour predictions.

Multiple-site arrays exhibit better predictability. Single-hour SEE is .16,

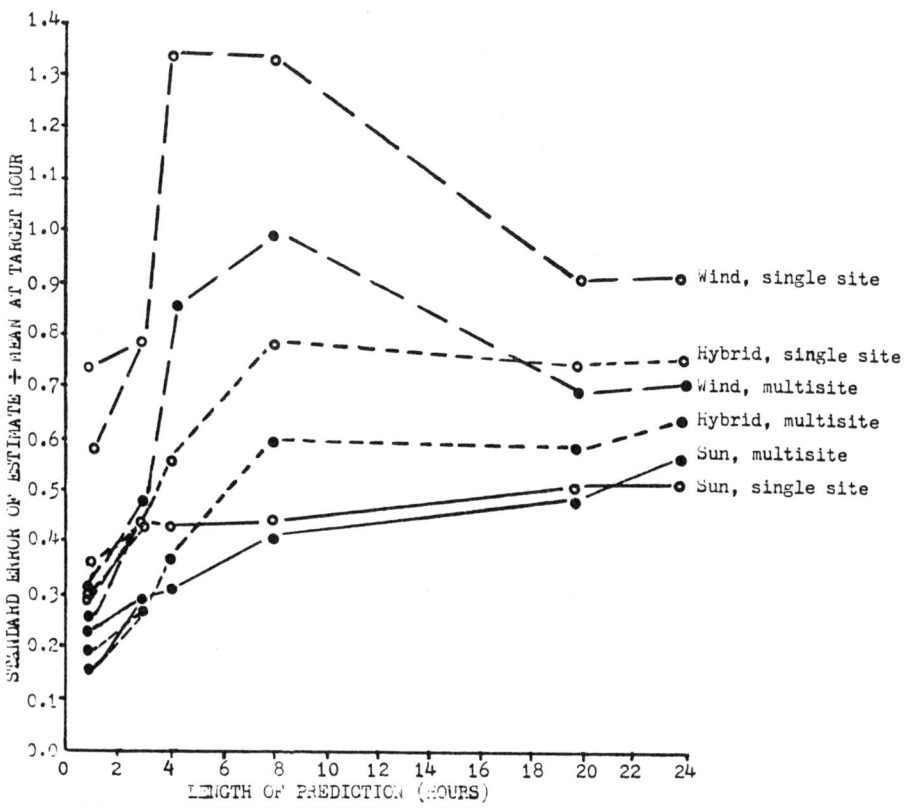

	TARGET HOUR = 1 P.M.			TARGET HOUR = 4 P.M.			
Length of Prediction	1	4	20	1	3	8	24
Solar Cells							
Single Site	.30	.45	.50	.29	.44	.40	.44
Multisite	.16	.31	.49	.23	.30	.42	.57
Wind Machines							
Single Site	.69	1.3	1.0	.59	.78	1.4	.92
Multisite	.30	.86	.70	.27	.45	1.0	.73
Hybrid Wind/Sun							
Single Site	.33	.61	.72	.35	.43	.80	.81
Multisite	.17	.38	.58	.20	.28	.60	.64

FIGURE 6. Predictability of Renewable Electrical Production (graph and table). Length of Prediction (Hours) versus Standard Error of Estimate ÷ Mean at Target Hour. Higher S.E.E./Mean indicates less predictability. Sun systems are most predictable. Source: James Kahn, "Decentralized Renewable Electricity Generation: Utility System Considerations and Connecticut Reliability Study" (masters thesis, Wesleyan University, 1981).

3-to-4-hour predictions are as good as single-hour predictions for the single site, and longer predictions are about the same as for single sites.

Wind machines exhibit less predictability. The 1-hour prediction has an SEE of .64 on the average; the 3-hour SEE is .78. Four and 8-hour SEEs average 1.35, indicating almost no predictability, while 20- and 24-hour SEEs average a slightly lower .96.

Multiple-site arrays are significantly more predictable. Single-hour SEE averages .29; 3-hour SEE is .45. Four- and 8-hour SEEs (.86 and 1.0) are the worst; 20- and 24-hour SEEs average .72.

Hybrid sun/wind systems are less predictable than solar cells and more predictable than wind machines, but they are much closer to sun systems. One-hour predictions have, on the average, an SEE of .34, several-hour predictions .52, and long predictions .76 or higher. The multiple-site array is almost as good as the solar system multiple-site array. The SEE for a 1-hour prediction is .19, for several-hour predictions .33, and .6 for long predictions. The small sacrifice in predictability may be offset by other advantages.

Persistence of High and Low Power Production
This measure models the persistence of variation of real DRET energy output from mean output. Storage level is used as an indicator. Average daily demand is assumed equal to average daily DRET output for each month, so that energy output variations accumulate in storage, creating reserves or deficits. The exact expression of the measure is: capacity of storage required to cover the largest difference between high and low storage levels, where capacity is measured in days of yearlong average DRET energy output. The use of average output in particular implies that evaluation of persistence and its implications for storage use should be based on the amount of conventional electricity displaced, not the conventional capacity displaced. Doing the latter requires knowledge of how much conventional capacity is displaced per unit of installed renewable systems. Figure 7 summarizes the findings.

Solar cells require relatively little storage to bridge highs and lows. The average requirement for a single site is 3.2 days, with a range of 1.5 to 5.3 days in the four months. Multiple-site arrays have a lower requirement: 2.6 days with a range of 1.2 to 5.5. (The increase in the high end of the range is probably due to certain untested sites individually requiring a higher level of storage.)

Wind machines require more storage. The yearlong average is 4.5 days, with a range of 3.2 to 7.0 days. Multiple-site arrays are not much better, in contrast to findings for solar cells. The average is 4.3 days, with a range of 3.5 to 6.0.

Hybrid sun/wind systems are better than either wind or sun systems alone, particularly with respect to keeping down the maximum end of the months' range. A single site has an average storage need of 3.1 days, with a range of 1.7 to 4.3. Multiple sites average 2.6 days of storage, with a range of 1.6 to 4.3 days.

Hybrid sun/wind/hydro systems in multiple sites are clearly the most preferable configuration with regard to storage measures. Average storage need is

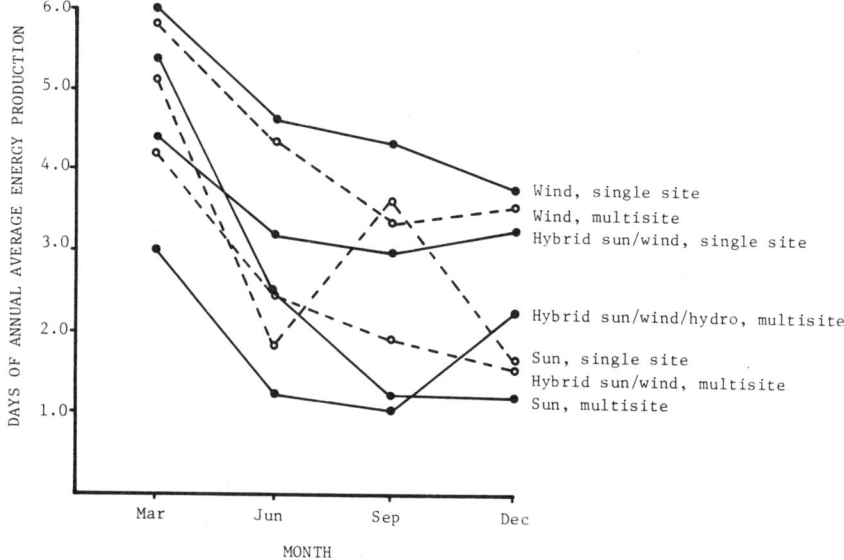

FIGURE 7. Storage Measure of Persistence of High and Low Renewable Electrical Production (graph and table). Days of Annual Average Energy Production, Storage Needed to Cover Maximum Fluctuation of Energy Output. Lower days is preferable; Hybrid sun/wind/hydro best for all but December. Source: James Kahn, "Decentralized Renewable Electricity Generation: Utility System Considerations and Connecticut Reliability Study" (masters thesis, Wesleyan University, 1981).

only 2.0 days, with a range of 1.2 to 3.2. In Connecticut there is insufficient indigenous hydro potential to create a large-scale hybrid of this type.

Storage measures have an application that other measures reported in this article do not share: they are useful in designing nonelectric renewable energy systems. Storage for thermal energy is cheaper than storage for electrical energy but still a major cost. Hour-to-hour variations are therefore not a concern, whereas day-to-day variations are the subject of careful analysis and planning. Considerable money might be saved if optimal combinations of renewable energy sources are designed for thermal systems. This may also affect electrical load-following requirements if electricity is used to back up renewable thermal systems.

The findings presented here suggest an important role for integrating sources and sites. Multiple-site arrays usually exhibit significant statistical superiority over single sites. Sometimes these advantages are pronounced, such as for predictability and persistence of power in the case of solar cells and for power distribution in the case of wind machines.

Hybrid and solar systems show traits much preferable to wind systems. They have similar characteristics, with each being preferable for different measures. Solar technologies are better for predictability and short-time-period power distributions. Hybrid systems are better for storage, longer-time-period distribution, and consistency of annual power production.

With hybrid systems, the problems of wind systems are corrected with little adverse impact on sun system traits, and sometimes with improvement. Sun systems alone make sense, wind systems less so.

Certain limitations suggest a careful interpretation of the results presented here.

1. The sites used are not optimal for wind speeds, and are therefore not prime wind machine locations. This problem, caused by the limitations of available data, is common for studies set in the less windy parts of the country, such as the inland Northeast. Findings are assumed applicable to excellent wind locations, as least in terms of general trends. The results for sun power are not so limited, as sun is fairly consistent throughout a region.

2. Few sites were studied. There are three combined wind and sun sites, one wind site with equally good data, and two wind sites with less frequent data collection. This total of six fails to indicate directly what could be expected with hundreds of sites in a large statewide renewable electric energy system; it is a better model of several widely dispersed energy farms. Extrapolation to many sites can be done by assuming that the benefits of adding a DRET site will be greatest in a system with few sites, and less pronounced but parallel effects can be expected from the addition of a site to a system with many sites. That is, diminishing returns can be expected as the number of sites increases. An important question, left unanswered in this study, concerns the limits of the desirable multiple-site effects: what are the exact traits of the best possible DRET configuration?

3. A statistically odd distribution of power from wind results from the technical traits of wind machines (cut-in and rated wind speeds).[11] In particular, power at zero and at maximum is more common than power near the mean. Therefore, the widely used statistical measures for a normal bell-shaped curve are sometimes inapplicable, and other techniques are used.

RECOMMENDATIONS

The state should encourage solar cell use before wind machine use, and encourage wind machines increasingly as solar cell systems go into operation.

The state should require electric utility companies to consider renewable energy systems as an integrated whole for computation of capacity credits for DRETs and resultant back-up electricity rates. The credits would then be accurate and also more favorable to renewable systems.

As renewables contribute a significant percentage of electrical production, the state should establish rate structures encouraging electrical use at times that make most effective use of renewable sources and minimize costs for conventional power.

The federal government should continue research on storage technologies relevant to the integration of renewable electrical technologies into electric systems.

The state should investigate the division of electrical demand into two categories. One would be the high-reliability type now always used for demands that must not be interrupted. It would be priced at a premium. Demand in the other category would tolerate supply interruptions of short or even daylong duration. This category would have lower prices because of saved capital and operating costs realized by less strict supply and demand matching. It could include freezers, water heaters (short-duration interruptions), and clothes washing machines (longer interruptions, with warning). Other demands could fit into this second category as well. The interruptions might even be limited to certain times of day (perhaps to peak demand times, or to peak generating hours for renewable sources). Such a division could help accommodate the unavoidable fluctuations in renewable-source electrical power.

NOTES

1. Studies that examine only multiple-site arrays include: JBF Scientific Corporation, *Wind Energy Systems: Application to Regional Utilities* (Washington, D.C.: U.S. Department of Energy, May 1979); C. G. Justus and A. S. Mikhail, "Energy Statistics for Large Wind Turbine Arrays," *Wind Engineering* 2 (1978):184–202. J. P. Molly, "Balancing Power Supply from Wind Energy Converting Systems," *Wind Engineering* 1 (1977):57–66. Studies that examine only single-site hybrids include: T. S. Jayadev, J. Henderson, and C. Bingham, "Conversion System Overview Assessment. Volume II: Solar-Wind Hybrid Systems" (Golden, CO: Solar Energy Research Institute, August 1979).

2. Studies that examine days and months include: Jayadev et al., "Conversion System Overview Assessment"; James E. Arnold, "On the Correlation between Daily Amounts of Solar and Wind Energy and Monthly Trends of the Two Energy Sources," *Proceedings of the 1978 Meeting of the American Section of the International Solar Energy Society* (1978), pp. 19–36 to 19–40; John W. Andrews, "Energy-Storage Requirements Reduced in Coupled Wind-Solar Generating Systems," *Solar Energy* 18 (1976):73–74. Studies that look at individual hours for single-source systems include: Justus and Mikhail, "Energy Statistics for Large Wind Turbine Arrays"; Bent Sørensen, "On the Fluctuating Power Genera-

tion of Large Wind Energy Converters with and without Storage Facilities," *Solar Energy* 20 (1978):321–31.

3. Andrews ("Energy-Storage Requirements") uses Blue Hill, Massachusetts, and New York City as sites for wind and sun, respectively. Ghazi Darkazalli and Jon McGowan, "Analytical Performance and Economic Evaluation of Residential Wind or Wind and Solar Heating Systems," *Proceedings of the 1978 Meeting of the American Section of the International Solar Energy Society* (1978), pp. 24–20 to 24–24. Darkazalli and McGowan use 1971 Hartford wind data and 1958 Blue Hill sun data.

4. One example of predictability analysis is Lennart Larsson, "Large-Scale Introduction of Wind Power Stations in the Swedish Grid: A Simulation Study," *Wind Engineering* 2 (1978):221–23.

5. The Northeast Utilities Environmental Data Acquisition Network (EDAN) was set up to collect wind and temperature data according to Nuclear Regulatory Commission regulations; solar data collection was added at NU's initiative.

6. Local Climatological Data sheets, available from U.S. Department of Commerce, National Climatic Center, Federal Building, Asheville, NC 28801. Also available in some government depository libraries, such as the Connecticut State Library in Hartford. Other types of records are also available, all for a fee.

7. Data are from the Connecticut Department of Environmental Protection Air Monitoring Division, State Office Building, Hartford, CT.

8. *Coastal Zone Wind Energy Study* (Washington, D.C.: U.S. Department of Energy, 1978). Data were acquired through personal communications with researchers from the project.

9. United States Geological Survey, *Water Resources Data for Connecticut* (water years 1976 and 1977).

10. Northeast Utilities Forecasting Section, personal communications.

11. The assumptions and mathematical methods used to simulate DRET energy production follow. Solar cells are assumed tilted at 31° to the horizontal, facing south, with no concentration of solar rays. Horizontal insolation data are converted to output power with the following steps. First, direct and diffuse components of sunlight are separated using a mathematical correlation of *total horizontal insolation recorded divided by total horizontal insolation above atmosphere* with *percentage of total that is direct*. The particular equations used were developed by Eldon Boes of Sandia Laboratories in Albuquerque, New Mexico. Second, direct and diffuse light incident on the solar cell is calculated with trigonometric equations accounting for the tilt of the earth and of the collectors with respect to the sun, and by reflectance properties of the cell coatings. Third, conversion to electricity is based on cell efficiency (assumed 12%) and electrical equipment efficiency (assumed 98%). Wind machines are standardized to a 200-foot hub height, with 75-foot radius blades. The cut-in velocity (at which power is first produced) is set at 35% of each site's 200-foot average wind speed, with speed extrapolated and interpolated to 200 feet using special equations (power laws) with site-specific coefficients. Rated velocity (at which the machine reaches maximum power) is set at 1.6 times average speed, and cut-out velocity (at which the blades face sideways to the wind so they do not turn, and no power is produced) is set at three times average velocity. Rated power is computed with an equation that uses atmospheric pressure, efficiency of conversion from wind to electricity, blade diameter, and rated velocity. Equations derived from these speeds and powers are used to compute instantaneous power based on wind speed data (Justus and Mikhail, "Energy Statistics for Large Wind Turbine Arrays"). To simplify comparison and combination of sites with different wind speeds, sites with particularly low speeds have their speeds increased by a constant multiplier to increase the average to 11.2 miles per hour. Hydropower plants are assumed to have a 40% capacity factor (produce 40% of the power they would if running full speed all the time) and to operate with the "run of the river" (no storage). Electrical output is based on stream flow, the height that water falls, and various equipment

INTEGRATING DECENTRALIZED TECHNOLOGIES

efficiencies (New England River Basins Commission, *Potential for Hydropower Development in New England* [Somerville, MA: NERBC, 1980]).

12. For example: General Electric, *Requirements Assessment of Photovoltaic Power Plants in Electric Utility Systems. Vol. 2: Technical Report* (Palo Alto: EPRI, June 1978).

13. For example, Justus and Mikhail, "Energy Statistics for Large Wind Turbine Arrays"; JBF Scientific Corporation, *Wind Energy Systems*.

DAVID A. HUETTNER

CHAPTER 9 | *Restructuring the Electric Utility Industry: A Modest Proposal*

Huettner argues that the efficiency of the existing electric industry may be improved by allowing the development of alternative production technologies, without the constraints of the monopolistic forces in the industry. Some economic modeling, Huettner says, suggests that the overall system will become more efficient in this free-enterprise scenario. If utilities controlled only the management of the industry and the production sector were relatively open, then additional flexibility and responsiveness could be created.

For example, some large power plants take so long to construct that they may be found unnecessary halfway through the project owing to a decrease in expected demand. Several smaller systems might have gone on line more quickly and would certainly have decreased the chances of such mismanagement through overcapacity.

To test the economics of the alternatives and coincidentally to improve the efficiency of the existing electric generation industry, Huettner suggests that the old free-enterprise system may be the most innovative planning concept we need to employ.

THE NEED FOR REFORM

The problem of providing the United States with an adequate supply of power at minimum social cost is a large one, but even this perspective is not large enough. Our national problems must be considered in the context of world energy economics. Increasing oil and gas imports, purchases of hydroelectric power from Canada and uranium enrichment services from Russia, and coal exports provide direct links between the United States and world energy economies. Worldwide energy consumption is increasing exponentially. It has been projected that fossil fuels, which provide 85 percent or more of the energy we use, will approach exhaustion in two centuries for coal and in much less time for oil and gas at current consumption levels. Increasing concern with the environmental impact of power production methods adds another dimension to the already complex process of choosing among alternate forms of power and alternate means of producing this power, and determining the desired level of total power production and its allocation among industrial, commercial, and residential uses.

It is likely that electricity will play a pivotal role in America's energy future because our largest remaining energy resources (thorium, solar radiation, wind, small hydro, and other renewable sources as well as coal) can be used to generate electricity, once environmental constraints are considered. The organization, performance, and flexibility of electric utilities are therefore of vital importance to all Americans. Although there are several determinants of current public policies toward this industry, the natural monopoly concept is the most important and the one upon which the range of public policy options is most dependent.

Briefly, the natural monopoly concept states that when unit costs decline continuously with increased plant and firm size, inexorable cost pressures force dominance by a single firm—a natural monopoly. In these cases the consumer is best served by granting a single firm a geographic monopoly and by regulating prices so that the cost savings are shared with the consumer and the natural monopoly is allowed a fair return on investment. Competition is contrary to the consumer's interest since it will lead to small-scale plants and firms having needlessly high costs and prices.

Given these choices and the widespread acceptance of the natural monopoly argument, all fifty states have elected to regulate the electric utility industry. Two problems have arisen, however. The first is that regulation has not been an effective substitute for competition. The second is that technological changes and the growth of geographic market sizes have invalidated the applicability of the natural monopoly concept.

Regulation has been an ineffective substitute for competition for at least three reasons.[1] First, companies, regulatory commissions, and even regulatory lag have not produced sufficient incentives for superior managerial performance, and regulation has given utilities, in effect, a cost-plus contract. Second, because regulation is by nature negative and backward-looking, regulatory commissions cannot and do not require firms to innovate, to improve, to reach certain standards of efficiency or optimal size, or to otherwise behave as if they faced substantial competition. Finally, regulation typically takes the organization of the regulated industry as given and can result in protection of the regulated firms and even discourage progress. In spite of the best thinking of economists and the well-intentioned efforts of utility commissions, regulation remains static, passive, and generally unimaginative. The discipline of the market is to be preferred to these efforts and the scope of electric utility regulation should be reduced.

The natural monopoly concept has become less and less applicable to the electric utility industry for at least two reasons. First, the development of long-range transmission technology has progressively increased the distance over which electricity can be transported and has enlarged marketing areas. Second, intensive per capita growth of electricity use and increased organization have created more and more geographic markets capable of supporting more than one optimally sized company. Several recent studies have concluded that current plant and firm sizes in this industry are larger than necessary to achieve economic efficiency and have failed to support the natural monopoly argument.[2]

As the appropriate technology, scale and costs of generating units, generating systems, and transmission equipment change over time, so must the location and type of plants and the number of enterprises. Unfortunately, regulatory constraints have retarded these changes and have not allowed new or existing firms to enter the market areas of existing regulated utilities. The importance of electric energy in our economy is too great to allow the efficient use of present technologies and the development and adoption of new technologies to be restricted by a small number of utilities, government institutions, and restraints developed decades ago.

A PROGRAM FOR REFORM

In addition to the inefficiencies and costs resulting from the present organization of the electric power industry, particularly the lack of competitive stimulus, the consumer must also bear the direct costs of federal, state, and local regulatory agencies. The lack of adequate incentives for utility managements and the need for a more dynamic, flexible industry structure require major reforms in the organization of the industry. While I recognize the technical necessity for the coordination of generation, transmission, and distribution, I am not convinced that all these functions need to be performed by one vertically integrated, regulated monopoly in each geographic area. Indeed, differing scale considerations may frequently call for generation and transmission facilities considerably larger than those needed for efficient distribution.

Deregulation of all facets of the industry is now neither feasible nor appropriate, but several factors suggest that the generating sector should be totally deregulated. First, as I argued above, the traditional natural monopoly arguments are invalid for the generating sector. Second, most long-run problems facing the electric power industry are in generation, not in distribution or transmission. Potential fuel shortages over the next few decades will make it imperative that the most efficient plant scale and technology be employed, particularly with respect to fuel consumption and environmental impact. For this reason the generating sector must be organized to encourage and allow maximum responsiveness to changing scale, technology, fuel availability, and pollution requirements. Changes in the organization of the generating sector will, of necessity, require some restructuring of the transmission and distribution sectors as well. The following paragraphs outline a deregulation program designed to inject competition and flexibility into the generating sector while improving the overall performance of the entire industry.[3]

Distribution

As a first step in this program, I advocate the complete separation of the distribution sector from generation and transmission. It has long been accepted that there are few economies of scale in electric power distribution over the normal range of sizes.[4] It has also been recognized that wasteful duplication of facilities would occur if competition were allowed. There is therefore definite validity in treating firms in this sector as individual natural monopolists and in providing an appropriate regulatory framework. Under my proposal, distribution companies (1) would be regulated; (2) would be allowed a fair return on their investment; (3) would be allowed no ownership interest in generating or transmission companies; and (4) would purchase their electricity by soliciting competitive bids from independent generating companies. Where desired, a municipal or cooperative firm could provide its own nonprofit distribution company, but it too would be required to buy power through competitive bidding.[5] While distribution companies would have a geographic monopoly for commercial and residential customers, I see no reason to exclude competition among them for industrial customers.

Generation

As a second step, I propose the complete deregulation of the generating sector of the electric utility industry and its complete divorcement from distribution. Independent electricity producers would sell their output on long-term contracts (five-year contracts, for example) to private, cooperative, or municipal distribution systems and would be barred from having any ownership interest in a distribution system or a transmission company. The natural monopoly argument for electric power generation has for many years been based, in large measure, on the significant economies of scale of steam-generating plants and interconnected generating systems. The argument is clearly invalid for interconnected generating systems since the economies of scale of an interconnected system can be obtained by sufficient coordination of its activities—common ownership of all the facilities involved is not necessary.

Insofar as system economies depend on plant economies of scale, the natural monopoly argument is that an unregulated competitive generating sector would not provide generating plants of sufficient size to achieve significant plant economies of scale. A simple recounting of the evidence presented in recent studies,[6] however, indicates that the natural monopoly argument is also invalid for generating plants for two reasons. First, most of the potential economies of scale can be obtained by generating plants and firms of only moderate size. Second, the present regulatory framework retards adoption of optimal plant sizes whereas a competitive framework would promote their adoption. Since my proposal would not necessarily block mergers or artificially restrict market areas to prescribed geographic regions, generating companies would have more flexibility to adopt plant sizes appropriate to the current and future population densities and geographic distances involved.

Over the next thirty years the demand for power will at least quadruple and population shifts will continue. The proposed deregulation framework would allow generating companies more flexibility and would foster a more rapid response to these dynamic structural pressures than would the current regulatory framework. Under a deregulated system, the constant threat of entry in the next competitive round of bidding should keep firms efficient even where only a single firm holds contracts in a region. Moreover, changing technologies will certainly require generating firms to adjust in order to keep their costs competitive. Competition in bidding will mandate such flexibility, particularly in regard to plant size. In a system where major long-term contracts for the provision of power are written every five years (let us say), the bid price should always be close to the marginal cost of providing the power with the most efficient plant scale and technology. While a winning bidder may be the sole supplier of electricity to a given distribution company, this monopoly is at a price determined in competition and is temporary for the period of the contract.[7] Strong incentives for continued efficiencies are therefore still present.

Transmission

Providing an adequate and properly coordinated system of transmission in the proposed system is perhaps the most difficult problem. The natural monopoly argument for transmission is easily verified by the tremendous and seemingly inexhaustible economies of scale reported in the engineering literature.[8] The argument is, in general, valid and suggests that transmission should be divorced from both distribution and generation. The transmission network should serve as a contract carrier in the provision of its service and should make the transportation of electrical energy available to any and all firms at or near long-run incremental cost.[9] The regulation and control of this important network ought to be vested in the Federal Energy Regulatory Commission (FERC), formerly the Federal Power Commission, or some similar federal agency. The problems of regulating electric transmission do not seem vastly different in principle from those involved in regulating a natural gas pipeline. Because of this, I propose that many of the principles of the Natural Gas Act of 1938 be applied to transmission companies. In particular,[10]

1. Transmission companies should be contract carriers having no financial interest in any generation or distribution company (and vice versa).
2. Transmission companies should be required to file rate schedules with the FERC and to change them only with prior FERC approval.
3. The FERC should be authorized to set just and reasonable rates and to eliminate undue preferences. To assist in rate determination, the FERC should be empowered to prescribe accounting methods and to ascertain the actual legitimate cost of the transmission network.
4. The FERC may order transmission companies to extend their facilities and to make physical connections with other transmission companies, generation companies, or local distribution companies if it finds that no undue burden is placed on the transmission company. FERC approval of

voluntary extension of interstate facilities and abandonment of them should be necessary.
5. The FERC may prescribe service areas within which transmission companies can extend facilities without approval, but the company should not be protected from invasion by another transmission company if the FERC finds this in the public interest.
6. The FERC should prescribe reserve requirements and other standards to ensure reliability of service. These standards should include both insurance and liability plans for damages suffered in a system failure.
7. The FERC may set minimum capacity requirements for transmission facilities to assure that reserve capacity is adequate to meet the needs of all firms using the facilities.
8. The FERC should not be empowered to regulate electricity prices at either the generating plant or the final consumer level.

The basic change that these proposals would make in the present method of regulating transmission networks is to prevent each transmission company from having a monopoly over a specified geographic area. Limited competition could be introduced if the FERC deemed it to be in the public interest—presumably in those cases where economies of scale are not an overriding concern. These proposals are based on a recognition that coordination and not common ownership is responsible for the benefits of a modern, interconnected power system. Indeed, the coordination of the present transmission system is left to a large degree to the independent negotiations of individual utilities.

EVALUATION

In this evaluation I seek to bring together the problems of the industry and the remedy proposed above. I shall also address what I see as some of the principal objections to this program. The major strengths of the proposed reforms are:

1. Quality of service will improve.
2. Location of plants will tend to be more nearly optimal than at present.
3. Joint ventures between utilities and the temptation to collude in other areas will decrease.
4. Access to power will be eased for municipal plants and small cooperative and private utilities.
5. The cost and difficulty of regulation will be greatly reduced.
6. There will be great incentives for managerial efficiency.
7. Plant size can be more nearly optimal than under the present system.
8. Dynamic flexibility will be introduced in technology, plant size, plant location, and transmission techniques. It will in turn make prices and costs closer to the marginal ideal over time.

I will now discuss each of these advantages and evaluate some of the difficulties that may arise if these changes are carried out.

At present the consumer has little control over the quality of service, and few regulatory agencies deem it appropriate to consider service complaints in rate-case proceedings.[11] The only penalty a company suffers for a blackout is the loss of revenue from sales. Brownouts (voltage reductions) result in virtually no penalty even though consumers complain of burned-out motors in heavy appliances (air conditioners, refrigerators, and so on). Although the proposed changes would not solve all these problems, there are opportunities for improvement. For example, generation and transmission companies would be selling power to distribution companies under contract, and poor quality of service could result in suits for violation of contract. Contracts for electricity from variable sources would reflect that limitation. Indeed, these contracts would probably include penalty clauses for blackouts and brownouts caused by the generation and transmission companies and could even include the posting of performance bonds.[12] Frequent service interruptions might even result in the loss of contracts. The funds raised from these penalties could be used to lower the rates charged to consumers. Since unregulated generating companies can incur losses as well as profits, they should seek to minimize outages both to maintain their reputations and to avoid economic penalties. Reserve margins should also be improved, particularly for many small municipal and cooperative distribution systems. Better maintenance and easier scheduling of planned outages may also result from the greater extent and coordination of interconnections.

At present, plant location is largely restricted to the geographic area of the distribution company. This results in many plants being constructed in areas where costs of fuel, construction, and pollution control are excessive. Under the proposed system, transmission facilities and generating plants would not be artificially confined to a particular geographic region; hence there is every reason to expect that plants will be located so as to minimize the true long-run costs of construction, operation, and transmission.

Joint ventures used to achieve optimal plant size and location under the present arrangement run the risk of promoting collusion in other, less beneficial areas. The pricing of industrial sales to customers who may be mobile in response to differing electricity costs would be a logical area for such agreements. These joint ventures also give the utilities a chance to coordinate their cases before state and federal regulators. If the system proposed here were adopted, there would be no need for joint ventures among distribution companies. Some consolidation of the present structure of the generation industry would probably be induced in order to take advantage of available economies of scale, but competitive bidding for long-term supply contracts at regular intervals and the potential for antitrust action should keep the generating sector close to the competitive price.

The present ties between generation and distribution have made it difficult for some municipal and cooperative utilities to obtain power at reasonable prices.[13] Under the proposed reform there would be no competition between the generating sector and the distribution sector and therefore no reason the municipal plants would not be supplied under the same terms and conditions as private distribution companies. Indeed, this may make the establishment and expansion of municipal distribution companies more likely.

At present, significant resources of both the utility companies and the states are devoted to determining a fair value of generation properties and providing for a reasonable return thereon.[14] Under the proposed system, the free competition in power generation would obviate the need for anyone to be concerned in decision making with the "original cost" of properties or the fair return for them. Investors owning generating facilities would face exactly the same risks and rewards as do other private investors.

Managers of electric utilities today have little incentive to reduce costs or to make profitable investments. Regulatory commissions allow utilities a "fair return" on any investments concerning technologies, fuels, and operating methods. For example, managers may alternatively build with the latest technology to satisfy their engineers' pride or with old technologies with which they are more familiar and comfortable. They may choose to build facilities that might be uniquely economical to a monopoly because of regulations and special treatment but that would not be economical in the marketplace. Most dangerous of all, the utilities may be truly indifferent and may seek the quickest and easiest solution to problems as they become manifest. Since generating firms would be allowed to fall into bankruptcy if they made poor decisions or were operated inefficiently, there is some reason to expect that their managements and stockholders would continue to take affirmative action to keep them operating in the most efficient manner possible.

Under the proposed system the managers of the generating companies would have every incentive to choose the best technology, locate it in the optimal location, and build it the most efficient size. Those failing to meet these objectives should incur losses or lose subsequent bids. Several recent studies[15] have estimated that there are substantial unrealized economies of scale in the generation of electric power, particularly among the smaller generating systems accounting for one-third of the U.S. generating capacity. This has resulted, at least in part, from many firms' being too small to build plants of efficient size. Under the proposed program, the generating companies would have to make the most effective use of present techniques, given present fuel and transmission costs, in order to win and keep their contracts with the distribution companies. These pressures, and greater flexibility in plant location decisions and mergers, should produce a more efficient distribution of plant sizes under current technology.

Since bids for long-term supply contracts would be let by the distribution companies about every five years on the average, the price of electricity should respond more rapidly to changes in the costs of production.[16] Whether these cost changes resulted from changes in capital equipment costs, plant size, technology, operating techniques, or fuel costs, they would be implemented as soon as proved economical and their benefits would be passed directly to the distribution companies and presumably to the public. Even within the contract period the bidders would have to estimate future changes in costs in order to arrive at the winning and profitable bid. This long-run approach to pricing is in stark contrast to the test-year, imbedded cost approach used to reflect today's costs and estimates of tomorrow's costs rather than yesterday's. Furthermore, firms engaged in the generation of electricity would have a constant inducement to explore the economies of various plant sizes and techniques as well as an incentive to promote the development and implementation of transmission technologies, particularly those technologies that would allow them to bid a lower delivered price and to locate at a better site with respect to fuel and other resource requirements. The availability of new and lower cost techniques should be quickly reflected in newly let contracts.

Generating firms with old plants would be forced to adopt the new price in future bids as well as the new technology if it pushes cost below present variable costs. These pressures should introduce a greater degree of dynamic flexibility into the generation and transmission sectors and should produce, over time, a more efficient distribution of plant size, plant location, transmission and generation techniques, and prices.[17]

Six possible disadvantages of the changes I have proposed here are:

1. Temporary dislocations during the transition from the present regulatory framework to the proposed framework.
2. Loss of service caused by bankruptcy of a generating company.
3. The danger of preemptive monopoly.
4. The danger of operating at less than optimum thermodynamic and economic levels because of uncoordinated decision making.
5. The possibility of increased financial and operating risks.
6. The problem of ensuring system reliability in the absence of common ownership.

The problems of temporary dislocations during a transition period should not override the long-run advantages of the proposed changes. Certainly adequate safeguards can be found to protect the rights of consumers, stockholders, managers, workers, and parties to existing contracts. While I favor an orderly transition period, I will here develop only a single example for transition.

The least disruptive method for making the transition would be to require only new power needs to be supplied initially through the contractual bidding system. Existing capacity could continue in operation until it became uneconomical or fully depreciated. With electricity demand growing at 7 percent per year, it would not take long for the bulk of generation to be outside the control of present distribution companies. This process could be accelerated by providing incentives for existing utilities to sell off their present plant units.[18]

The occasional bankruptcy of a generating company and the resulting unfulfilled service contracts are not merely possible under the proposed reforms—they are expected. The posting of performance bonds and the usual rules of contract law should provide distribution companies with adequate financial safeguards in such situations, but these safeguards will not protect the consumer from loss of service while the contract is up for new bidding. If the termination of service is due merely to a financial disaster and not a physical disaster to a generating company, then the problem is simply the lack of a legal entity financially able to operate the generating plant. Certainly some legal or insurance program could provide for the temporary operation of bankrupt generating companies. If service is terminated because of a physical disaster, the problem is the lack of generating capacity and the solution will be much the same as it is under the present regulatory framework. It may even be better owing to the existence of a more highly interconnected generating system with sufficient reserve capacity and possibly greater diversity of sources. Since our proposed reforms advocate the centralized regulation of transmission by the FPC, they would also include the power to set appropriate reserve requirements.

The danger of preemptive monopoly arises if a generating company builds sufficient capacity to "preempt" a remote geographic area for itself. Such an action could discourage competition, but this result is not necessary for several reasons. First, distant generating companies with excess capacity can compete in the bidding by means of the transmission network. Second, a distant bidder could assume the preemptive monopolist will bid high and could offer a reasonable bid based on the belief that he will win and have an opportunity to lease the idle plant of the losing preemptive monopolist. Bidding strategies could be a highly effective check on the preemptive monopolist. Finally, a small or remote locality could form its own municipal generating company to bid against the monopolist on the contracts offered to the local distribution company.[19] It should also be noted that preemptive monopoly power, when it exists, is limited to the term of a contract and subject to great risk when the contract is up for bidding. Furthermore, in most cases the problem should be no greater than that already encountered in other unregulated industries such as cement, cans, and steel, where transportation costs are also a high proportion of total costs.

While the proposed deregulation of the generating sector creates the possibility of preemptive power, it also unleashes competitive forces that in most cases are powerful enough to offset this power. To further reduce the possible dangers of regional or local monopoly power at the generating level, I propose that a clear and specific policy be followed to discourage such concentration except where demonstrated scale economies are such as to necessitate increased concentration. As a generating firm's share of the generation market in any FERC region or other well-defined market area grows, the firm's cost must decrease for it to win additional contracts in the competitive bidding. Figure 1 illustrates one possible set of such trade-offs between the risk of monopoly power and the cost savings offered to distribution firms. In this example, once a generating firm's market share of a FERC region reaches 30 percent, the generating firm could not win an additional contract in that region unless its bid was at least 5 percent lower than the bid of a smaller competitor. As the dominant generating firm's share of the market increased further, so would the penalty under which it was bidding (and vice versa).[20] This policy would limit the profitability of the dominant firm unless that profitability was the result of efficiency or scale economies based on present or prospective technologies or plant sizes.

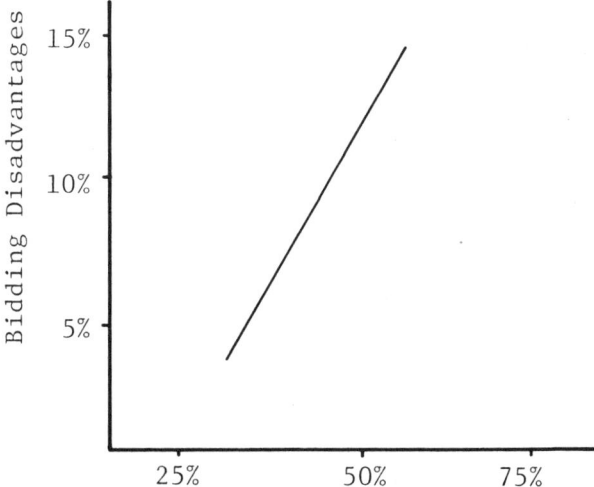

FIGURE 1. Bidding Disadvantage of Dominant Generating Firms.

The fourth disadvantage of the reforms proposed here is that uncoordinated decision making could lead to less than optimum efficiency since generation, transmission, and distribution would be performed by separate companies. It should be noted that this same criticism can be made of the present vertically integrated, separately owned system of electric power companies, since they are neither managed nor optimized as a single system. Although considerable regional cooperation is already occurring in both system planning and operational control, the major factor blocking further cooperation is a matter of sovereignty: no manager or technical head can be expected to appreciate being told what to do by a group of people outside his company.[21] I believe that centralized regulation of transmission by the FERC will enhance the chances for the most efficient planning, location, and operation of the U.S. power network.

Another objection to the proposed reforms is the danger of increased financial and operating risks. One advantage of vertical integration is the opportunity for improved planning of a sequence of interdependent activities. This improved planning can lower costs by lengthening production runs, allowing better scheduling and automation of material flow, permitting consideration of alternatives, and so on. Since these benefits would be the result of coordination and not of common ownership, I see no particular reason to assume that operating risks or uncertainties would be greater under the proposed reforms. To the extent that financial risks are increased by the absence of a regulatory framework that virtually eliminates financial failures, it is likely that the borrowing cost of electric power companies will rise. There is no reason for continuing this "subsidy" when the price of electricity can represent the true social costs. Other high-fixed-cost industries such as cement manufacturing have survived without regulatory "guarantees" of a fair rate of return. This change will also force electricity to compete more directly with other fuels in the marketplace.

In addition, it is possible that the large financial requirements for a single generating plant ($160 million for a 400-MW fossil plant and $480 million for an 800-MW nuclear plant) could forestall entry into the market by new generating companies. The long gestation period for plant construction and the threat of construction delays are additional obstacles. It is not likely that any of these would preclude entry since a generating company should be able to finance new construction based on its contract to sell the output to a distribution company. Risks of construction delays could be hedged by purchasing insurance covering such an occurrence.[22]

The sixth objection to the proposed reforms is the problem of ensuring system reliability in the absence of common ownership. Since all equipment in a system is exposed to damage or temporary shutdown if one piece of equipment fails, it is proper to require that appropriate measures be taken to ensure system reliability. I have here proposed that the FERC be empowered to prescribe reserve requirements and other standards to ensure reliability of service. These standards should be an effective substitute for the pecuniary incentives existing under common ownership. Certainly many existing power pools provide reliable service without common ownership of all equipment. While this proposal does not detail the exact nature of these complex reliability standards, it is difficult to accept the argument that common ownership of equipment is required to ensure system integrity.

One advantage of the trade-offs illustrated in figure 1 is that they are not identified with any particular plant size or technology. Therefore, in a dynamic sense, they would encourage innovation and even growth by dominant generating firms as long as socially desirable cost reductions were achieved. These trade-offs should be established (and revised) and administered by the Justice Department with the advice of the FERC. If a future technology were to increase optimum scale to the point where monopoly at the generating level became a serious problem, it would of course be possible to impose regulation on the generating sector once again. As long as regulation did not extend to restricting new entry it could be applied consistently with the objective of keeping price in line with long-run marginal costs. The knowledge that regulation could be imposed, that the cost disadvantage in bidding will grow with market share, and that new entry could occur should induce generating firms to act in such a manner that the reintroduction of regulation would not be necessary.

Finally, opponents of the proposal could argue that the separation of distribution, generation, and transmission could reduce the administrative size of power companies and hamper their efficiency. Studies of economies of scale in administrative costs of electric power companies conclude that there are either constant returns to scale or else only minor cost reductions for firms larger than 1,000 MW.[23] Furthermore, there is no reason to expect that the average size of power companies will decline if my proposal is adopted. While some of the largest vertically integrated power companies may decline in size, it is very probable that the smaller companies will have an opportunity to increase in size once artificial geographic restrictions are removed.

Adoption of this deregulation proposal is, of course, a decision to be made by each state individually. In making this decision, each state must consider the economic effects of deregulation, particularly if states competing with it for industrial expansion and jobs do not choose to deregulate. While the price and availability of electricity are but two of the many factors to be considered in industrial location decisions, deregulation of the generating sector will undoubtedly affect both by removing regulatory inefficiencies affecting generation and by giving industrial customers more opportunities to buy power in wholesale markets or from alternative distribution companies. While industrial rates will still be regulated by either a state commission or the FERC, the increased purchasing options granted industrial customers by our proposals would, in effect, end cross-subsidies between industrial customers and other classes of users.

Adoption of our proposals could either raise or lower prices paid by industrial users in the short run depending on the preexisting circumstances in each state (that is, the extent of regulatory inefficiencies and cross-subsidies between classes of users). In the long run, however, prices paid by industrial customers should be lower in states adopting our proposals since the marginal costs of generation should be lower and large industrial customers should be able to purchase power at prices based on marginal costs.

If my proposals were adopted, rates would rise as long as construction costs of generating units continued to inflate. Electricity prices would need to rise to cover the construction costs of new generating units and owners of existing units would experience windfall profits. Currently, most state regulatory commissions and the FERC use original cost to value the rate base. The windfall profits resulting from inflation in construction costs are, in effect, allocated to consumers by regulatory commissions in the form of rates lower than replacement costs. Current regulatory practices, therefore, discourage energy conservation by pricing electricity below replacement cost. My proposal would end this, at least during periods of construction cost inflation. Since the windfall profits would accrue to owners of generating plants, I recommend that the deregulated generating sector be subject to a windfall profits tax during the ten years following deregulation.

NOTES

1. For a more complete review of regulatory problems, see Harry M. Trebing, ed., *Performance under Regulation* (East Lansing, MI: Michigan State University Press, 1968); Ben W. Lewis, "Emphasis and Misemphasis in Regulatory Policy," in William G. Shepard and Thomas G. Gies, eds., *Utility Regulation: New Directions in Theory and Policy* (New York: Random House, 1968); Alfred E. Kahn, *The Economics of Regulation* (New York: John Wiley & Sons, 1970 and 1971); William Capron, ed., *Technical Change in the Regulated Industries* (Washington, D.C.: Brookings Institution, 1972).

2. See, e.g., D. A. Huettner, *Plant Size, Technological Change and Investment Requirements* (New York: Praeger, 1974); and D. A. Huettner and J. H. Landon, "Electric Util-

ities: Economies and Diseconomies of Scale," *Southern Economic Journal* 44 (1978):883–912.

3. A similar proposal for reform, at least with regard to vertical disintegration, now appears in a draft bill prepared by the staff of the Senate Subcommittee on Antitrust and Monopoly. For an alternative reform proposal, see Leonard W. Weiss, "Antitrust in the Electric Power Industry," in Almarin and Phillips, eds., *Promoting Competition in Regulated Industries* (Washington, D.C.: Brookings Institution, 1975).

4. The cost per unit of capacity of distribution equipment such as transformers, control equipment, busses, and so on, decreases with the capacity of the equipment. Since fixed costs are 85 percent of total distribution costs, there are potential economies of scale in distribution, but they are heavily dependent on the load density and other characteristics of the area served. For a more complete discussion see The Distribution Technical Advisory Committee for the National Power Survey, *The Distribution of Electric Power* (Washington, D.C.: Federal Power Commission, June 1969), pp. 82–92, 151, and 185.

5. Exceptions to this rule would certainly be required in certain remote areas.

6. See, e.g., D. A. Huettner, "Shifts of Long Run Average Cost Curves: Theoretical and Managerial Implications," *Omega, the International Journal of Management Science* 1 (1974): 421–50; and D. A. Huettner, "Scale, Cost and Environmental Pressures: Nuclear and Fossil Steam Plants," in Bela Gold, ed., *Technological Change: Economics, Management and Environment* (Oxford: Pergamon Press, 1975).

7. The loss of a bid by the owner of an existing generating plant should not result in the wasteful construction of a new generating plant by the winning bidder, for several reasons. The winning bidder may have excess capacity. If he does not have excess capacity, then he and the losing bidder could negotiate a lease for the existing plant as long as the existing plant is at least marginally efficient. Wasteful duplication of plants is not a necessary consequence of the bidding process.

8. See, e.g., The Transmission Technical Advisory Committee for the National Power Survey, *The Transmission of Electric Power* (Washington, D.C.: Federal Power Commission, February 1971), pp. 20–25; and The Advisory Committee on Underground Transmission, *Underground Power Transmission*, (Washington, D.C.: Federal Power Commission, April 1966), pp. B20–B25.

9. This proposal is merely an extension of the Supreme Court decision in the Otter Tail Power Co. case (1964).

10. These proposals are patterned on the discussion of the Natural Gas Act presented by Merle Fainsod, Lincoln Gordon, and Joseph C. Palamountain in *Government and the American Economy* (New York: W. W. Norton, 1959), pp. 666–67. Note that I do not favor the regulation of electricity prices by the FERC as it regulated natural gas prices.

11. See, e.g., *Elyria Telephone Company v. P.U.C.O.*, 158 Ohio State 441 (1953). Recently, however, the state regulatory commissions of both Florida and California have withheld rate increases to induce regulated companies to develop satisfactory service quality standards. See United Telephone Company of Florida, Docket no. 9012–TP, Order no. 4360 (74 PUR 3rd), 1968, and General Telephone of California, Case no. 8749 (80 PUR 3rd), 1969.

12. These penalties and the size of the performance bond could be set by the FERC or else determined by negotiations between the contracting parties.

13. See especially Richard Hellman, *Government Competition in the Electric Utility Industry* (New York: Praeger, 1972), and John T. Miller, Jr., "A Needed Reform of the Organization and Regulation of the Interstate Electric Power Industry," *Fordham Law Review*, vol. 38, p. 635.

14. A reasonable guess as to the costs to the companies of regulatory proceedings would be two to three times the public expenditures.

15. See, e.g., William R. Hughes, "Scale Frontiers in Electric Power," in W. M. Capron, ed., *Technological Change in Regulated Industries* (Washington, D.C.: Brookings Institution, 1971); D. A. Huettner and J. H. Landon, "Restructuring the Electric Utility Indus-

try: An Economic Analysis," Case Western Reserve University, Research Program in Industrial Economics, Working Paper no. 43, 1973; and S. Ling, *Economics of Scale in the Steam-Electric Power Generating Industry* (Amsterdam: North Holland, 1964).

16. Of course, some fuel-adjustment clause may be required to compensate the generating firm for sudden and unanticipated increases or decreases in costs. We suggest, however, that such clauses offer less than full compensation in an amount great enough to maintain incentive to hold costs down.

17. J. H. Landon, "Changing Technology and Optimal Industrial Structure," in Bela Gold, ed., *Technological Change: Economics, Management and Environment* (Oxford: Pergamon Press, 1975).

18. Safeguards may be necessary to prevent the existing vertically integrated distribution companies from lowering resale prices in areas where competition is likely and raising prices in other areas to compensate for the loss of revenues.

19. The tax-free status of municipal generating companies that compete with private taxpaying companies raises issues that deserve attention once generation is deregulated.

20. The percentage cost to the consumer of this bidding penalty would generally be much smaller than the penalty itself because only a small percentage of the total power used by consumers would be affected. For example, assume that a firm controlling 50 percent of the market submits a bid for an expiring contract on the remaining 50 percent of the market. The maximum additional cost penalty to be borne by the consumer is a 10-percent cost increase for half of his power or a 5-percent penalty on his total power bill.

21. P. Sporn, *Social Organization of Electric Power Supply* (Cambridge, MA: MIT Press, 1971), p. 145.

22. Indeed it is likely that some generating companies would purposely overbuild capacity in order to sell insurance to surrounding companies.

23. J. McNulty, "Administrative Costs and Scale of Operation in the U.S. Electric Power Industry," *Journal of Industrial Economics* 5 (1956):30–43; and D. A. Huettner and J. H. Landon, "Electric Utilities: Economies and Diseconomies of Scale," paper presented at the Western Economic Association Annual Meeting, June 1974.

EDWARD THOMPSON III

CHAPTER 10 | *Cogeneration and Small Power Production: Some Intergovernmental Policy Concerns*

When Congress formulated a comprehensive energy policy in 1978, it recognized the potential significance of alternative energy sources in electrical generation. The Public Utilities Regulatory Policies Act (PURPA) of that year is testimony to the impact that these energy sources can have. The PURPA legislation, in effect, recognized that the existing utility regulatory process was discouraging the use of these alternatives and identified the utilities themselves as an important partner in the decentralization process particularly in the grid-interface technology discussed by Morris in chapter 5 and Sørensen in chapter 6.

Thompson introduces the PURPA legislation from the perspective of regulation through intergovernmental cooperation. Such strong federal legislation may sometimes create confusion about the roles of states, local governments, and regulatory agencies. Thompson attempts to clarify these aspects of PURPA and cites examples of its successes and failures.

When the United States enjoyed abundant and inexpensive energy supplies, energy policy was not a major concern of citizens or government. However, the price and supply shocks that have occurred since 1973 have altered this situation. Citizens, administrators, and government officials are now focusing concentrated attention on energy issues.[1] Since energy policy issues represent a new concern for the political system, there has been a relatively rapid evolution in energy relations between federal and state governments.

INTERGOVERNMENTAL ELECTRIC ENERGY RELATIONS

The question of intergovernmental energy relations has recently been addressed by Joan B. Aron in an article suggesting that such interactions are largely in the pre-1937 "conflict" phase.[2] This term, originally used by Deil S. Wright, describes federal–state government relations in this phase as "adversary and antagonistic patterns of interaction."[3] Given this conflict, federal and state energy policy relations are "characterized by official disputes over defining jurisdictional boundaries and allocating governmental responsibilities."[4]

In examining these conflicts, Aron suggests that states have become dominant in the formulation of energy and electric energy policy. For example, she notes that "the states are posing an immediate challenge to federal authority in those cases where both levels of government seek to establish control over the same or similar activities."[5] In a later passage she makes the dominant role of state governments more explicit:

> The states are unlikely to accept severe federal restraints on their organizational or regulatory activities. Given the importance the states attach to electric energy and to energy issues generally, it would be unreasonable to expect them, on their own volition, to relinquish authority to a national interest.... Although federal officials and utilities may be increasingly concerned over the new state assertiveness, it is unlikely that the states will back off unless their legislative judgements and determinations are overturned by Congress or the courts.[6]

THE SHIFT TO COOPERATIVE ELECTRIC ENERGY RELATIONS

In contrast to the Aron article, this chapter suggests that while federal–state energy relations are marked by both conflict and cooperation, they now appear to be largely "cooperative." This term, also used by Deil S. Wright, characterizes federal–state relations as collaborative and cooperative. In large part this cooperation reflects the perceptions policy makers have about the "seriousness" of a problem confronting the political system. For example, intergovernmental relations assumed a cooperative form from 1933 to 1953, a period of major economic and international crisis. Hence, Wright notes that "the prime elements of national concern during those two decades were the alleviation of widespread economic distress and response to international threats. It seems logical and natural that internal and external challenges to national survival would bring us closer together."[7] I will suggest that the energy crisis has had a somewhat similar effect on important federal–state policymakers since 1973. Although there are some exceptions, it appears that the energy crisis is promoting greater intergovernmental electric energy policy cooperation. Accordingly, federal–state energy relations are moving away from their conflict phase. Moreover, this movement has been compressed into a relatively small time period as a result of the nature of the country's electric energy problems.

Although this chapter suggests that intergovernmental energy relations are largely cooperative, some conflict still surrounds many energy issues. The coal severance tax and energy resource development in the West provide examples of how the interests of the federal and state governments can conflict. On the whole, however, there appears to be less conflict than previously.

In part, intergovernmental cooperation stems from the characteristics and problems of the electric utility industry. Until the early 1970s the relative costs and increases in the price of electricity were moderate. However, electrical costs have increased sharply since the beginning of the decade. For example, the average retail price of electricity rose 162 percent from January 1972 to February 1980, while the Consumer Price Index increased only 92 percent.[8] The steep increase in electricity prices prompted widespread criticism of industry practices that appeared to contribute to inefficiency and higher costs. Citizens began to organize around such utility issues as declining block rates, automatic fuel-adjustment clauses, construction work in progress, and lifeline rates.[9]

THE PUBLIC UTILITIES REGULATORY POLICIES ACT

The national government responded to increased public concern over the environmental, cost, health, and safety consequences of utility actions by passing the Public Utilities Regulatory Policies Act (PURPA) of 1978. This chapter examines selected portions of Title II of the act to see how they affected intergovernmental electric energy regulations. However, it is important to note that PURPA is only one of five related laws passed as the National Energy Act (NEA) of 1978.[10] As the first comprehensive national energy policy statement, NEA established quantitative energy conservation and production goals to be achieved by 1985.[11]

Of PURPA's six titles,[12] Title I is especially important. It proposes six utility rate-making and five utility policy standards for consideration by state regulatory commissions and large nonregulated utilities.[13] In broad terms, Title I has three purposes. First, it encourages the conservation of energy supplied by electric utilities. Second, it seeks optimal efficiency in the use of facilities and resources by utilities. Third, it encourages equitable rates to consumers.[14] To achieve these purposes the act requires that regulators conduct hearings on each standard, that citizen intervenors be accorded certain rights in those hearings, and that regulators issue written determinations of their policy decisions on each standard.

Although many states have not yet implemented their rate-making and policy standards,[15] Title I has already had a major impact on the nation's utilities. For example, in its General Order no. 33 the New Mexico Public Service Commission (NMPSC) requires state utilities to file detailed reports on their efforts to achieve PURPA's goals. The commission notes that its decisions under Title I have been largely congruent with the "basic PURPA standards by identifying viable alternatives available to the utilities . . . for achieving the energy conservation policy ends adopted by the Commission. . . . It is the Commission's hope and intent that through a flexible and continuing monitoring and evaluation process it can cause all of the necessary aspects of conservation to become an integral part of day to day thought and planning processes of our utilities as well as their customers."[16]

The decision of the NMPSC suggests that states may be adopting a cooperative attitude toward the provisions of PURPA Title I. PURPA does not establish a penalty if a regulatory body or nonregulated utility fails to meet scheduled deadlines. Instead, it requires only that consideration and determinations be made under the standards established by the law. However, evidence is increasing that state commissions are complying with Title I. For example, a nationwide survey conducted by the National Association of Regulatory Utility Commissioners (NARUC) found "that the state commissions will, with few exceptions, discharge their obligations to consider and determine the PURPA standards in compliance with the statute and within prescribed time limits."[17] Should regulatory commisions continue to adopt this position, Title I of PURPA could promote significant rate-making and policy changes in the operation of the nation's utilities, suggesting that intergovernmental conflict over electric energy policy has decreased.

TITLE II OF PURPA

However great the potential impact of Title I, Title II of PURPA may prove of more enduring significance.[18] In general terms, it eliminates many of the obstacles that have hampered the production of electricity by industrial cogenerators and independent small power producers.[19] Until the present, "self-generators" have confronted three major problems in their efforts to produce electricity. First, the utilities have not provided equitable rates to self-generators for their sale of nonutility-generated power. Second, the utilities have charged self-generators inequitable rates for the power they require when their generating systems are down. Finally, cogenerators and small power producers have confronted a maze of laws and regulations governing the production and sale of electrical power.

Although these three problems have discouraged the production of power by self-generators, cogeneration and small power production are proven technologies that could contribute significantly to the policy goals of NEA and PURPA. At present, 27 percent of West Germany's electric power is produced by cogeneration. Similarly, the Energy Project at the Harvard Business School found that "over twenty percent of total industrial energy use could be saved in the United States through cogeneration."[20] It also concluded that renewable-energy small power production technologies "could have significant impact on the traditional distributors of energy, the utilities, which could find themselves partially bypassed, leading to a leveling off or actual decline in demand for their services."[21]

Not only is self-generated power potentially useful in helping to meet the goals of NEA and PURPA, it can also increase the decentralization of U.S. energy production and reduce pollution.[22] However, for the laws to succeed in accomplishing Congress's goals, advocates of small power production and cogeneration will have to participate effectively at the state level in implementing the provisions of Title II. Meaningful citizen participation is essential because most utilities will likely resist the competition self-generators will bring to the energy market.

Although PURPA required states and unregulated utilities to conduct hearings on their plans to implement Title II regulations and to report back to the

Federal Energy Regulatory Commission (FERC) by March 21, 1981, many states have continued the decision-making process after this date. Additional hearings are probable because "in most states little more than general principles and trial rate schedules are expected to emerge from the initial proceedings."[23] Even when rates have been established, small power producers and cogenerators will still be able to negotiate with regulators for more favorable rates. Given these considerations, the ongoing interpretation of Title II promises to be a dynamic and extended process.

The above discussion suggests that the PURPA Title II process will give self-generators an opportunity to influence state-level decision makers if state officials adhere to the goals established by Congress in NEA and PURPA. By focusing on two sections of Title II (210 and 201), I will attempt to ascertain whether federal–state electric energy interactions are cooperative. I will pay particular attention to the implications of section 210, which, in broad terms, seeks to increase the economic viability of alternative power sources by enabling small power producers and cogenerators to purchase back-up power from utilities and sell any surplus power they generate back to the utility grid. In addition, section 210 "orders FERC to design rules to exempt cogenerators and small power producers from all or part of the Federal Power Act, the Public Utility Holding Company Act, and state utility regulation, in order to encourage the development of these facilities."[24] This exemption eliminates many of the procedural and regulatory barriers that have impeded the development of these resources. Section 201 of Title II provides the operating and efficiency standards used to determine whether a self-generator is entitled to the benefits provided by PURPA. Besides operating characteristics, the section also establishes ownership requirements for self-generators seeking PURPA benefits. The success of these two sections in meeting the electric energy policy goals of Congress is largely being determined, as noted previously, by actions at the state level. I shall now examine two procedural issues—sales to utilities and interconnection—addressed by sections 201 and 210.

SALES TO UTILITIES

Perhaps the most critical issue confronting small power producers and cogenerators is that of rates. The rates that qualifying facilities[25] receive from selling power to utilities will in large part determine their financial viability. This issue is also important for establishing the degree of cooperation between federal and state officials. If state officials establish low rates, they can thwart the intention of Congress. However, rates that are adequate to ensure the operation of qualifying facilities will promote the energy goals of NEA and PURPA. It is therefore critical that rates provide adequate compensation to small power producers and cogenerators.

In 1980 the FERC issued rules providing that qualifying facilities should be paid rates that are "just and reasonable" and based on the "avoided costs" of the purchasing utility. Considerable controversy surrounds the meaning of "just and reasonable." However, avoided costs are more easily defined as the incremental capacity costs and/or energy costs an electric utility would incur if it were unable to purchase power from a qualifying small power producer or cogenerator. In other words, avoided costs are fuel or plant capacity expenses that a utility escapes when it purchases power from a qualifying facility. Such a purchase frees a utility from being forced either to generate extra power itself or to purchase that power from another utility. The controversy surrounding "avoided cost" purchases is whether power from a qualifying facility displaces only fuel from existing plants or whether it displaces new plant capacity requirements. If the facility displaces capacity requirements, it will lower the large capital requirements of the utility. Qualifying facilities that lower fuel and capacity costs should therefore receive higher payments.

In determining whether rates are just and reasonable, a number of factors must be considered. Obviously, qualifying facilities are concerned about the price paid for generated power. In addition, they require some assurance that the rate established by state officials for sales to utilities will be relatively stable. This is important because qualifying facilities can represent large economic investments. Therefore they require regulatory assurance of rate stability in order to become financially attractive.

To a great extent the above concerns have been answered by the New Hampshire Public Utilities Commission (NHPUC). The NHPUC has granted small power producers 8.2¢ per kWh for the capacity costs they allow utilities to avoid. Small power producers that allow utilities to avoid only energy costs receive a lower rate of 7.7¢ per kWh. In addition, the rates set by the commission are applicable "to all small power producers presently operating qualifying facilities and to all small power producers who activate qualifying facilities between the date of this order and the date of initial generation at Seabrook for the life of the qualifying facilities."[26] With this decision, the NHPUC has initiated payment of the rates discussed above to all currently operating small power producers. This action provides these facilities with the economic assurance they require to become viable.

The New Hampshire rates are "generic" in the sense that they apply to all transactions between small power producers and utilities.[27] Commissions can also establish generic rates for different types of power and for self-generators of different sizes. For example, the Idaho Public Utilities Commission (IPUC) has established two tier rates for firm and nonfirm power. Firm power is power that qualifying facilities make available at all times during a contract, except during forced outages and scheduled maintenance shutdowns. With firm power, a utility has legally enforceable guarantees that a given amount of power generated by a qualifying facility will be delivered. It can thus avoid construction or the purchase of power from another utility. By way of contrast, nonfirm power is delivered by the producer at his option. Since the qualifying facility has this option, nonfirm power cannot be used by utilities in calculating their capacity needs. However, nonfirm power does allow the purchasing utility to lower its overall costs.

The distinction between the two types of power is used by the IPUC. The commission determined that qualifying facilities delivering firm power are eligible to receive capacity costs and fuel costs. Accordingly, Idaho facilities receive an average price of 5.46¢ per kWh. By contrast, the IPUC granted qualifying facilities delivering nonfirm power only 3.82¢ per kWh. However, it should be noted that the IPUC did award nonfirm power producers an additional "small amount in consideration of system capacity benefits."[28] This is especially important because utilities do derive capacity benefits from the cumulative effect of having many nonfirm qualifying facilities.[29] A similar stance has been taken by the California Public Utilities Commission (CPUC). It has stated that nonfirm qualifying facilities are eligible for rate payments amounting to half of the "full 25–30 year full avoided costs" of new plants operating at 100 percent capacity factors.

State commissions have also decided that payments to qualifying facilities should reflect any future rapid shifts in the price of oil. The price of oil has been considerably more volatile than the price of other fuels such as coal and uranium. This precludes any long-term price averaging when determining the energy cost payments to qualifying facilities. For example, if costs are calculated yearly, fuel cost increases occurring after the yearly calculation will not be reflected for several months. This procedure can shortchange facilities of deserved fuel costs when prices are increasing rapidly. The CPUC has noted that "rapid and successive increases in oil prices require that . . . actual avoided cost be reflected as accurately and rapidly as possible in its energy payments."[30] Accordingly, it has ordered that costs be calculated quarterly.

The above discussion suggests that decisions issued by state commissions on pricing factors[31] are promoting the development of small power production and cogeneration facilities. The pricing decisions reached in California, Idaho, and New Hampshire, as well as in Connecticut,[32] New York,[33] and other states conform closely to the FERC regulations. For example, the state-by-state survey conducted by the NARUC concluded that "a number of commissions have already complied with the Section 210 requirements, by establishing rates for purchases of electricity from qualifying cogenerators and small power producers, by establishing rates for back-up power, [and] by establishing rules for interconnection."[34] Again, this suggests that states have adopted a cooperative attitude toward federal electric energy regulation.

INTERCONNECTION

"Interconnection" is the process whereby qualifying small power producers and cogenerators are physically linked to an electric utility system. This linkage is necessary to enable a qualifying facility to purchase or sell electricity to a utility.

Interconnection presents self-generators, federal and state officials, and utilities with several key policy questions. The answers that state-level decision makers provide will help determine the viability of qualifying facilities. Moreover, these answers will also help determine the degree of cooperation between federal and state officials.

Foremost is the question of "reasonable" costs. The FERC regulations state that

> "interconnection costs" by definition means, the *reasonable* costs of connection, switching, metering, transmission, distribution, safety provisions, and administrative costs incurred by the electric utility *directly* related to the installation and maintenance of the physical facilities necessary to permit interconnected operations with a qf, to the extent such costs are *in excess* of the corresponding costs which the electric utility would have incurred if it had not engaged in interconnected operations.[35] [Emphasis added]

In order to define what costs are "reasonable," regulators must first agree on what equipment and methods of operation provide for safe interconnections. However, opinions on interconnection safety vary widely among qualifying facilities, equipment manufacturers, and utilities. Potential cogenerators and small power producers in Texas have already "expressed concerns about being charged for extraneous safety and interconnection equipment."[36] They note that if a utility charges exorbitant interconnection costs, it can prevent a facility from coming on line. Moreover, equipment manufacturers can naturally be expected to favor high equipment requirements.

To help clarify controversies in this area, the staff of the CPUC addressed the major issues surrounding reasonable costs. In its recommendations the staff argued that qualifying facilities are responsible for those costs "in excess of the corresponding costs which the utility would have incurred if it had not engaged in interconnected operations."[37] However, in making this recommendation, the CPUC staff provided qualifying facilities with several safeguards. First, it supported the standardization of reliability and safety requirements for interconnected equipment because those standards could provide guidelines that would simplify interconnection for the facilities. The staff reasoned that standardization of requirements would reduce the administrative and regulatory cost burdens confronting small power producers and cogenerators. Furthermore, it recommended that the guidelines vary according to the size of the generating facility. For example, small power producers were not to be held to the same standards as large industrial cogenerators. Finally, the CPUC staff supported guidelines that specified "functions that must be provided rather than specifying a list of equipment."[38] This approach lowers interconnection costs incurred by qualifying facilities by removing restrictions on their freedom to select different devices or practices to achieve reliability and safety.

The method of payment of interconnection costs is another important issue for small power producers and cogenerators. The burdens and expenses confronting qualifying facilities can be eased if flat interconnection fees are charged. The CPUC found that this practice can "eliminate the need for case-by-case negotiations between the future qualifying owners and the utility."[39] In addition, regulators can ease the financial burdens of qualifying facilities by timing the interconnection cost payments. If initial costs are high, the CPUC staff noted, they may prove prohibitive for some small power producers. Accordingly, innovative financing schemes that involve the utilities should be considered. The CPUC staff recommended that facilities be granted the option of either paying in advance the estimated net cost of installation or amortizing the cost over a period of years, possibly through payments on the monthly utility bill.

The findings presented here, suggesting that federal–state energy interactions have shifted to a more cooperative phase, are supported by early state decisions issued under PURPA.[40] On balance, these decisions advance the Title II goal of reducing obstacles to the production of electricity by qualifying self-generators. States are providing small power producers and cogenerators equitable rates for power sales, as well as reasonable interconnection costs. Moreover, these findings are similar to those of the NARUC. The NARUC national survey notes that, with few exceptions, the decisions made by state commissions are removing the "regulatory obstacles preventing the aggressive promotion of alternative sources of energy."[41]

Federal–state energy relations have thus been undergoing relatively rapid change since 1978. The passage of PURPA, NEA, and the Energy Security Act (ESA)[42] points to growing intergovernmental cooperation, if not federal preeminence, in developing energy policy. However, there are indications that these relations are still in a state of flux. At least two factors could slow or even reverse the trend toward federal preeminence.

First, it is possible that PURPA will be declared unconstitutional. In early 1981 the state of Mississippi, the Mississippi Public Service Commission (MPSC), and the Mississippi Power and Light Company (MP&L) brought suit concerning the constitutionality of PURPA. The suit *(State of Mississippi v. FERC and U.S. Department of Energy)* alleged that under PURPA Congress had exceeded its authority to regulate the functions of state governments. Moreover, the state argued that (1) PURPA's requirement that Mississippi officials consider federal standards and PURPA's granting of intervention rights to citizens and the secretary of the Department of Energy had impaired the MPSC's ability to function as a regulatory body; (2) Congress was forbidden to impose its choices upon the states regarding the conduct or structuring of state activities.

On February 19, 1981, the United States Southern District Court of Mississippi issued the first court decision on the constitutionality of Titles I, II, and III of PURPA. The court declared that, under the Commerce Clause, the federal government did not have the authority to assume control over the activities of public utilities. The decision also cited the Tenth Amendment to the Constitution, which provides that powers not delegated to the federal government are reserved to the states. Accordingly, the court declared that "the United States does not have the power or authority to impose its three standards under PURPA upon the state of Mississippi under the guise of providing the solution to the nation's energy problems."[43]

The FERC has declared its intention to appeal the decision. However, if the

Supreme Court eventually declares the first three titles of PURPA unconstitutional, the trend toward greater federal authority in electric energy policy could be reversed.

There are also indications that the federal government may retreat voluntarily from the Carter administration electric energy policies. Such a retreat would have a negative impact in states that are not well advanced in their PURPA proceedings or that do not have progressive commissions. However, the effect will not be as great in states such as California and Idaho, which do have progressive commissions. Nonetheless, a retreat by the federal government from its cogeneration and small power production policies would place it in conflict with the more progressive states. Ironically, therefore, the federal government may have started a cooperative process under PURPA that eventually will contribute to greater conflict.

NOTES

1. One recent survey found that the public is most hostile to energy price increases. It further noted that the incident at Three Mile Island has increased public concern over environmental, public health, and safety issues as they relate to nuclear power plants. In addition, a widespread sense of alienation and cynicism toward the government, local utilities, and the oil and gas industry has arisen. See Ronald D. Brunner and Weston E. Vivian, "Citizen Viewpoints on Energy Policy," *Policy Sciences* 12 (1980):147–70.

2. Joan B. Aron, "Intergovernmental Politics of Energy," *Policy Analysis* 5 (1979): 451–71.

3. Deil S. Wright, "Intergovernmental Relations: An Analytical Overview," *Annals of the American Academy of Political and Social Sciences* 416 (1974):6.

4. Aron, "Intergovernmental Politics of Energy," p. 452.

5. Ibid., p. 453.

6. Ibid., p. 469.

7. Wright, "Intergovernmental Relations," p. 7. The reader should be mindful that opinions vary on the nature, as well as the causes, of the current energy "crisis." In this regard see William Darity, Jr., Ronald Johnson, and Edward Thompson III, "Energy and Equity in a Broader Social and Economic Context" (Paper presented at the Brookings Institution, "High Energy Costs: Assessing the Burden," October 9, 1980).

8. Eunice S. Grier and George Grier, *Too Cold, Too Dark: Rising Energy Prices and Low-Income Households* (Washington, D.C.: Community Services Administration, 1980), p. 7. The Grier study demonstrates that electric utility practices have a regressive economic impact on low-income groups.

9. This is only a partial listing of the issues that have mobilized citizen opposition to the utility industry. Other groups, such as New Hampshire's Clamshell Alliance, have organized around the construction of nuclear plants. The Citizen Action League of California and Massachusetts Fair Share organized for general rate reform and lifeline rates. These terms and issues are more fully addressed in Richard E. Morgan, *The Rate Watcher's Guide* (Washington, D.C.: Environmental Action Foundation, 1980).

10. The National Energy Act is composed of the following public laws: P.L. 95–617, Public Utilities Regulatory Policies Act; P.L. 95–618, Energy Tax Act; P.L. 95–619, National Energy Conservation Act; P.L. 95–620, Powerplant and Industrial Fuel Use Act; P.L. 95–621, Natural Gas Policy Act. Among other things, the NEA supports coal conversion by utilities and industries using natural gas and oil; conservation loans and a consumer insulation program in which the utilities can play a role for homeowners and renters; federal

energy efficiency standards for major appliances that would preempt state standards; conservation for schools and other local institutions; loans for solar applications; and weatherization grants for low-income people. The provisions of the five laws demonstrate that the national government has decided to take a more active and leading role in promoting its energy goals.

11. In large part the various energy proposals developed by the Carter, Ford, and Nixon administrations were predicated on the assumption that the increasing world demand for oil threatened to outstrip the world's capacity to produce. There was also concern that domestic oil and gas supplies were in danger of being exhausted. Some analysts (Mobil Oil, the National Academy of Sciences, and the National Petroleum Council) have projected that domestic supplies will be depleted by the beginning of the next century. Although these projections are debatable, they did have an influence on the president and Congress. For example, the Carter energy plan outlined in the president's July 15, 1979, speech stressed the need to reduce oil imports and gas consumption. Moreover, the administration committed itself to a massive expenditure of federal funds and a broad relaxation of environmental regulations in its proposals for the Energy Security Corporation and the Energy Mobilization Board. Although these two proposals represented radical departures from previous policy, they were consistent with Carter's pledge to reduce national dependence on foreign oil by declaring it the "moral equivalent of war." This pledge, which quickly became known as part of the "MEOW" speech, demonstrates the seriousness with which the national government approached the issue.

12. The most widely contested portions of the act are those pertaining to the electric utility industry. Title I examines utility rate structures and policies; Title II provides incentives for cogeneration and small power production and contains provisions regulating wholesale rates; Title III requires state commissions to review the termination proceedings and advertising cost recovery policies of natural gas utilities and mandates the Department of Energy to study gas utility rate design issues; Title IV promotes small hydroelectric energy projects through a sponsored loan program; Title V addresses crude oil transportation systems; Title VI contains a variety of miscellaneous provisions. A more complete discussion of Title I of PURPA can be found in Alden Meyer, *A Ratepayers Guide to PURPA* (Washington, D.C.: Environmental Action Foundation, 1979).

13. The six rate-making standards in Title I are: cost of service; declining block rates; interruptible rates; load management; seasonal rates; and time of date rates. The five policy standards are: advertising; automatic adjustment clauses; information to consumers; master metering; and termination procedures.

14. U.S. Congress, Conference Report, Public Utility Regulatory Policies Act, 95th Cong., 2d sess., 1978, p. 5.

15. PURPA establishes different time limits for consideration of its rate and policy standards. The rate structure standards must be completed and a determination made on whether to implement them by November 1981. Consideration and determination of the regulatory policy standards were scheduled for completion by November 1980. Covered utilities without lifeline rates were scheduled to conduct hearings by November 1980. See Meyer, *A Ratepayer's Guide to PURPA*, p. 15.

16. Ad Hoc Committee on the National Energy Act, *State Commission Progress under the Public Utility Regulatory Policies Act of 1978* (Washington, D.C.: National Association of Regulatory Utility Commissioners, December 1980), pp. 6–7.

17. Ibid., p. 4.

18. This chapter focuses on sections 201 and 210 of PURPA. However, Title II also includes regulations concerning utility interconnections and wheeling; studies on power pooling; studies on wholesale rate-making by the Federal Energy Regulatory Commission; interlocking directorate information requirements; automatic adjustment clause study requirements; and requirements that the Department of Energy conduct a study of electric system reliability.

19. Cogeneration is the combined production of electricity and heat, steam, or some other form of useful energy by an industrial heating or cooling system. Small power production refers to the production of electricity from renewable energy sources, such as solar, wind, biomass, or waste.

20. Robert Stobaugh and Daniel Yergin, eds., *Energy Future: Report of the Energy Project at the Harvard Business School* (New York: Ballantine, 1980), p. 199. This view is supported by the Idaho Public Utilities Commission, which claims: "Two generations ago, American industry produced 25% of its own power needs. Today, European industries continue to enjoy production levels of that magnitude, whereas American industry has slipped to producing only 4% of its own power" (Idaho Public Utilities Commission, Order no. 15746, August 8, 1980, p. 3).

21. Stobaugh and Yergin, *Energy Future*, p. 234. This point has also been substantiated by the FERC. It estimates that by 1995 PURPA will induce the following development in qualifying technologies: industrial cogeneration, 3,400 MW; commercial cogeneration, 2,500 MW; wind generation, 1,900 MW; municipally owned solid waste (MSW) to electricity generation, 360 MW; small-scale hydropower generation, 3,500 MW. See Federal Energy Regulatory Commission, *Small Power Production and Cogeneration Facilities—Qualifying Status/Rates and Exemptions* (Washington, D.C., June 1980), p. VII–12. This passage might give the impression that utilities are "losers" because of the move to independent power production. However, alternative generation sources provide major benefits to utilities as well as to society. Among other things, alternative generation sources consume fuels more efficiently; diversify the utility's resource plan; increase system reliability and minimize risk; require smaller reserve margins; require less construction lead time; and have lower capital requirements. See California Public Utilities Commission, Decision no. 91109, December 19, 1979, pp. 13–14.

22. Among the environmental benefits attributable to PURPA are reduced stack gas emissions from coal combustion; reduced fugitive dust from coal handling and storage; reduced wastewater discharges from reduced cooling tower and boiler blowdowns; reductions in the generation of solid wastes; reduced air pollutants from gas turbine and combined cycle power plants; and a reduction in the risks associated with nuclear power plants. See FERC, *Small Power Production and Cogeneration Facilities*, pp. VII–30–33. Although small power production and cogeneration technologies have some negative environmental consequences, the FERC notes that "existing environmental and siting regulations can mitigate most of the potentially adverse environmental effects that may occur because of the Sections 201 and 210 regulations" (p. VII–33).

23. Alan Miller and Barrett Stambler, *Promoting Small Power Production* (Washington, D.C.: Environmental Action Foundation, 1981), p. 3.

24. Meyer, *A Ratepayer's Guide to PURPA*, p. 27.

25. The FERC has established requirements concerning the ownership and operating characteristics of self-generators. Facilities that qualify under the FERC regulations are entitled to benefit from the PURPA legislation. The 1980 regulations limit electric utilities or electric utility holding companies to a 50-percent equity interest in a qualifying facility. After specifying what constitutes a cogeneration facility, the regulations establish operating and efficiency standards for cogenerators. The small power production regulations limit the size of these facilities to 80 MW. In addition, small power producers must derive more than 75 percent of their total energy input from biomass, renewable resources, or waste. Finally, small power producers and cogenerators have the option of applying to FERC for certification as a qualifying facility or self-certifying their facility after providing FERC with information about their capacity, energy source, and location. A more complete explanation of this process can be found in David Silverstone, *PURPA Provisions on Cogeneration and Small Power Production* (Washington, D.C.: Environmental Action Foundation, 1980), pp. 4–8.

26. New Hampshire Public Utilities Commission, Decision 79–208, June 18, 1980, p. 12. It should be noted that energy and capacity payments vary between and within states

COGENERATION AND SMALL POWER PRODUCTION 211

because so many factors are involved in computing avoided costs. Nonetheless, it is possible to establish avoided costs that are "accurate," just, and reasonable.

27. No cogenerators appeared before the NHPUC when it made its decision. As a result, the commission established different rates and requirements for cogenerators.

28. Idaho Public Utilities Commission, Decision no. 16025, December 2, 1980, p. A–1.

29. This view is also supported by the Massachusetts Institute of Technology Energy Lab, which notes "that the contribution of intermittent sources (e.g., photovoltaic) can be aggregated to estimate their contribution to the utility system." See California Public Utilities Commission, *Final Staff Report on Cogeneration and Small Power Production Pricing Standards* (San Francisco, 1981), p. 85. It is also important to note that there are variations in reliability between different technologies. The FERC rules allow for this variation to be reflected in payments to qualifying facilities. The facilities with the highest reliability receive the largest payments.

30. California Public Utilities Commission, Decision no. 91109, December 19, 1979, p. 18.

31. This discussion has examined only some of the pricing issues being decided by states. Other factors, such as the age of a facility, are considered by state commissions when they issue regulatory decisions.

32. Connecticut Department of Public Utility Control, *Cogeneration and Small Power Production Facilities, Additional Rules and Regulations*, Rate 981, Docket no. 800404, December 22, 1980.

33. New York Public Service Commission, Case no. 27824, January 7, 1981.

34. NARUC, *State Commission Progress*, p. 17.

35. Federal Energy Regulatory Commission Rules, sec. 292.101(b) (7).

36. *Report of the 1980 Task Force on Cogeneration in Texas to the Public Utility Commission of Texas*, October 1980, p. 33.

37. CPUC, *Final Staff Report*, p. 12–3.

38. Ibid., p. 14–3.

39. Miller and Stambler, *Promoting Small Power Production*, p. 11.

40. For example, the CPUC fined Pacific Gas and Electric $7.2 million for having a "passive attitude" toward the development of cogeneration conservation projects. The CPUC made explicit mention of PURPA and the FERC cogeneration rules in its order. See California Public Utilities Commission, Decision 91107, December 19, 1979.

41. NARUC, *State Commission Progress*, p. 17.

42. U.S. Congress, Conference Report, *Energy Security Act*, 96th Cong., 2d sess., 1980, Report no. 96–824. The Energy Security Act represents the most recent federal effort to assume control over national energy policy. Several provisions in the ESA clearly demonstrate the preeminence of the federal government in intergovernmental energy relations. The act calls for massive expenditures for synthetic fuels, gasohol production, and other "biomass-energy" initiatives. Smaller expenditures are allocated for a Solar Energy and Conservation Bank and for a four-year "Municipal Waste Energy Development Plan." Perhaps the most controversial aspect of the act is its provision of up to $88 billion in financial guarantees for the synthetic fuels industry. The law also gives the Department of Energy and the Tennessee Valley Authority the power of "eminent domain" for use in acquiring state, private, and Indian lands and allows the utilities to participate in solar energy and conservation markets. On balance, the law greatly increases the power of the federal government.

43. *State of Mississippi v. FERC and U.S. Department of Energy*, United States District Court Southern District of Mississippi Civil Action no. J79–0212(c), p. 6. In issuing the decision Judge Harold Cox also stated that "the sovereign state of Mississippi is not a robot, or lackey which may be shuttled back and forth to suit the whim and caprice of the federal government" (p. 2). It should be obvious based on the decisions from other states that the position adopted by Mississippi is not the norm. For example, other states have justified cogeneration, small power production, and rate-making incentive decisions on

PURPA. The IPUC decided that "the general rate-making purposes set forth in Section 101 of PURPA—mainly, conservation, efficient use of resources and facilities by electric utilities, and equitable rates to consumers—are consistent with the statutory duty of this Commission to insure adequate electric service at just, reasonable and non-discriminatory rates" (Idaho Public Utilities Commission, Case no. P–300–18, March 6, 1981). In this decision the IPUC issued generic regulations for the six PURPA rate-making issues covered under Title I, justifying its decision by citing the provisions contained in PURPA.

PART III

Issues in Decentralization

DAVID A. HUETTNER

CHAPTER 11 | *Diseconomies of Scale*

This chapter summarizes a study that reviewed and disputed many traditional economic assumptions about the electric industry.[1] Huettner suggests that close scrutiny shows long-standing and accepted conclusions about economies of scale to be inaccurate, even by the standards of those economists and planners who accept them. The idea that "bigger might not be better" is explored in some depth. Huettner introduces analyses indicating that the natural monopoly concept does not entirely make sense in the electric utility industry, and he demonstrates that the methodology and assumptions of previous studies have significant shortcomings.

I. INTRODUCTION

For many decades the electric power industry has been regarded as one of the more efficient sectors of American industry. Many observers have cited improvements in the productivity of all inputs as evidence of the industry's rate of technological progress and ability to exploit economies of scale. Economic studies of both plant and firm-level economies have concentrated almost solely on the generating sector of the industry, but they have generally reached similar conclusions with some important disagreements about the source of the efficiencies (i.e., economies of scale, technological change or utilization of capacity) and the ranking of productivity improvements among the inputs. None of these studies, however, has ever identified any diseconomies of scale or even challenged the natural monopoly concept as applied to the electric power industry.[2]

The conclusions reached in this chapter are important because they highlight the shortcomings of previous studies of the electric power industry;[3] they extend to the firm level recent conculsions about scale diseconomies at the plant level; and, by questioning the natural monopoly concept as applied to the industry, they suggest that the range of public policy options vis-à-vis the industry should be broadened to include deregulation of some functions. While recent price and cost increases in the electric power industry have led some observers to speculate that scale economies may be nearing exhaustion, these speculations have not been documented in any systematic way or based on current or past studies. In fact, past studies (Barzel 1974; Cowing 1974; Dhrymes and Kruz 1964; Galatin 1968; Johnston 1960; Kirchmayer et al. 1955; Komiya 1962; Ling 1964; Lomax 1962; McNulty 1956; Nerlove 1968; Olson 1970) have consistently reported the existence of scale economies throughout the range of observations.

The first studies documenting scale exhaustion at the plant level (Huettner) appeared in 1974 and 1973 respectively. The first published study indicating scale exhaustion at the firm level (Christensen and Greene) appeared in 1976. These two researchers (hereinafter C and G) concluded that there were constant returns to scale in generation for firm sizes as low as 19.8 billion kWh (about 3,800 MW). However, this study has somewhat limited public policy implications for several reasons. First, it concentrated solely on costs directly allocable to generating and therefore did not address scale issues in other cost categories comprising 50 percent of total costs. Second, while its use of a translog cost function does not restrict the form of the production function and hence the elasticities of substitution of the inputs, it does restrict the consideration of other important differences among firms in the sample and could introduce specification errors. These important differences include variations among companies in the degree of capacity utilization; in the types of fuels used; in their reliance on nuclear, hydro, or gas turbine capacity and purchased power; and in regional demand patterns, peak demand, and construction types and costs. In addition, the procedure of summing the individual firms of a holding company and treating them as one entity assumes that the degree of integration is both high and constant across holding companies.[4] This procedure should shift the true scale curve to the right and flatten it out. Finally, C and G's use of accounting data on depreciation to measure capital (following Nerlove 1968) should introduce a scale-opposed bias, as has been noted in Huettner 1973.

In a real sense, a researcher studying scale issues is frequently forced to choose which type of specification error he prefers. For example, if he chooses a translog cost function as did C and G, then restrictions on the production function and elasticities of substitution among inputs are minimized but at the cost of not controlling for important differences among firms that could well affect the scale conclusions. If, on the other hand, the researcher chooses a more pedestrian cost function that allows him to use dummy variables and otherwise control for important differences among firms, he will reduce this type of specification error but increase errors associated with restrictions on the elasticity of substitution among inputs.

C and G have taken the former approach while in this chapter we have taken the latter and have also chosen to treat holding companies differently. In addition, we have elected to examine scale relationships in generation, transmission, distribution, administration, customer accounts, and sales. Finally, we have used an entirely different concept of cost function than is usually used in economics. Although this approach is necessary to capture the cost implications of meeting peaked electricity loads, it deserves some justification since it is a marked departure from traditional economic theory and previous studies of the electric power industry. Section II of this chapter will justify the use of this nontraditional cost concept.

Turning to another issue, it is worth noting that relationships observed at the plant level, particularly economies of scale, are often modified by interrelationships at higher levels of decision making, such as the firm and system level. The system level is perhaps the most natural level of analysis for this study, but at least two factors indicate that statistical work cannot be confined to systems. First, existing electric power systems take many forms varying from loose to tightly knit.[5] Second, there are few relevant data on systems. For these reasons the empirical analysis of this study will be based exclusively on the firm level. Firm-level relationships, however, can be predicted by taking the plant-level relationships observed in Huettner 1974 and 1973 and combining them with the system-level relationships described in the engineering literature (Ewald and Angland 1964). This analysis[6] provides some scale predictions or benchmarks against which the regression results summarized in section III can be compared and indicates that the long-run average cost curves for generating should be U shaped and that operating and fixed costs in generation should be a minimum for firm sizes of 2,000 MW and 3,000 MW respectively.

Section III summarizes and evaluates the results of a multivariate regression analysis assessing firm-level economies of scale in generation, transmission, distribution, and administration. Existing econometric studies cannot form the basis for this analysis because, as noted above, they have had several important shortcomings, have concentrated primarily on generation and excluded transmission and distribution, and have, with two exceptions, been based on data existing prior to 1959.

Two points should be noted before turning to the next section. First, 1971 data will be used in this study to rule out most of the effects of pollution-control equipment on scale relationships. Very few of the plants operating in 1971 employed pollution-control equipment, but since that date many plants have been retrofitted with such equipment or forced to burn fuels that they were not designed to burn.

The second point is that Averch-Johnson, or A-J, effects (i.e., a capital-intensity bias in equipment selection) will be ignored throughout this study. This position is justified for several reasons noted by Boyes (1976) and by others, including the capital constraints on the electric power industry; the existence or absence of an A-J effect depending on the model assumed; and the small or nonexistent A-J effects reported in empirical studies.

II. COST FUNCTIONS FOR PUBLIC UTILITIES

Traditional economic theory produces cost functions that have input prices $P_1, \ldots, P_n$ and the quantity output Q as arguments and a functional form determined by the production function as in Equation 1:

$$C = f(Q; P_1, \ldots, P_n)$$

These traditional functions are useful for characterizing operations where peak demands do not have to be met (i.e., order backlogs) or operations where output can be stored and peak demands met by sales from inventory. Under these conditions, since production costs are not heavily influenced by demand peaks or by the annual pattern of demand, we can assume that production takes place at an even rate or that output is an appropriate measure of the scale of production, as in Equation 1.

Public utilities (and some private-sector activities as well) generally provide services such as electricity, communications, or transportation, which cannot be stored. Because they are also usually required to meet demands for their services at all times of the day or year, they must build capacity capable of producing the quantity demanded at peak periods. Meeting a peaked load is clearly more expensive than meeting an even load even if the total quantity produced is the same. In addition, the quantity of output loses its meaning as an accurate index of scale, particularly in capital-intensive industries such as public utilities, where input relationships are determined by the level of planned activity—peak capacity requirements.

The above arguments suggest that costs and scale relationships in public utilities could be better measured if costs were expressed as a function (g) of peak capacity, K, the rate of utilization of peak capacity, U, and input prices in Equation 2.

$$C = g(K, U, P_1, \ldots, P_n)$$

Of course, $Q = KU$ but does not appear in Equation 2 since peak capacity and the pattern of annual demand relative to peak capacity are the factors determining scale and costs.

One issue not usually addressed in either cost or production function studies concerns the durability of capital. Public utilities are generally capital intensive, and the physical or economic life of capacity is an important consideration usually ignored by economic theory. Indeed, it is difficult to see how it could be incorporated into traditional production function analysis, although a start has been made at least from the cost side (Huettner 1973). We do not pretend to have the answer to this question, but we do explicitly recognize that we have skirted this issue whereas the study by Christensen and Greene does not.

Our method of skirting the issue is simply to separate operating and fixed costs and to express fixed costs in terms of dollars per kilowatt of capacity instead of dollars per unit of output (i.e., killowatt-hours). This procedure allows us to examine scale effects on fixed costs without making assumptions about economic life versus plant size and about the particular depreciation method to be used. For a more complete discussion of these issues see Huettner 1973.

Because this study is concerned with a particular type of public utility—electric utilities—we have elected to use the nontraditional cost concept described in Equation 2. We realize that this may give our cost function an "ad hoc" flavor to which some may object, but it should be remembered that our approach does allow testing of certain firm characteristics not otherwise examined

III. ECONOMIES OF SCALE FOR ELECTRIC UTILITIES

This section summarizes an empirical analysis of firm-level economies in generation, transmission, distribution, sales, customer accounts, and administration. The study did not attempt to disentangle cost advantages or disadvantages at the firm level from those arising at other levels of decision making but merely tested the degree to which large firms are associated with lower unit costs.[7] As noted earlier, the transmission and distribution functions have not been subjected to empirically based economic studies, but all the engineering-based studies, such as Federal Power Commission 1969, 1971, 1966, have reported substantial economies of scale in both fixed costs and operating costs. Of the four firm-level economic studies reviewed in Huettner 1974 and 1973, three reported substantial economies of scale in generation and the single study of administration costs concluded that there were constant returns to scale. In addition, all four were based on pre-1955 data.[8] Clearly our understanding of firm-level economies is fragmented, out-of-date, and based mainly on engineering relationships or economic studies concentrating on the generating function alone.

The regression analysis summarized was based on 1968 and 1971 data taken from the FPC. The sample was limited to the seventy-four electric utilities that sold to residential, commercial, and industrial customers; had fossil steam capacity comprising at least 80 percent of their total generating capacity; and generated at least 80 percent of their own power. Single-equation models were developed for each of the major cost categories comprising total operating expenses except for depreciation charges, which have been excluded from the analysis. The use of single-equation models is defended on the grounds that the data to develop a demand model for each utility are currently unavailable.[9] It should also be recalled that 1971 data were used in this section to rule out most effects of pollution-control equipment on scale relationships.

The major cost categories examined in this section include, for operating cost: production; transmission, distribution; administrative and general; customer accounts; and sales expense. Long-run average variable cost curves were estimated for each of these expense categories. In addition, scale relationships were adjusted and undepreciated fixed investments per kW of capacity were estimated for production.[10]

Two basic measures of firm size were employed in the regressions—company generating capacity in megawatts (MW) and total annual company sales in megawatt-hours (MWh). Generating capacity was used as the firm size measure for production, transmission, and distribution expenses since these are most closely related to annual sales, particularly for companies that purchase a significant portion of their power.

The square of the firm size variable was included in each regression to test for U-shaped long-run average cost curves. An additional test of the scale effect was also made by performing separate regressions on firms having generating capacity under 2,000 MW and those with generating capacity over 2,000 MW. Based on the literature reviewed in section I, we would expect to find long-run average variable cost curves declining throughout the entire range of observation. The system level of analysis referred to in section I, however, questioned this view and suggested that L-shaped or U-shaped cost curves are more likely to occur.

Utilization of generating capacity and its square were also included in the production, transmission, and distribution cost regressions to test for a U-shaped short-run average cost curve. Additional variables included in these regressions were input prices, regional dummy variables (costs were expected to be lowest in the South and highest in the Northeast), and holding company dummy variables (costs were generally expected to be lower for companies belonging to holding companies). Other variables specific to each cost category were also included in each regression.[11]

Economies of Scale

Table 1 presents a summary of the scale results. For five of the six operating cost categories and for the sum of all six categories (total operating expenses) there are diseconomies of scale beyond moderate firm sizes. Only sales expense, a minor component of total operating costs, exhibited scale economies across the entire range of observed firm sizes. Unit costs for fixed investment are available only for production plant, but again the results indicate that there are diseconomies of scale beyond moderate firm sizes. The results in table 1 indicate that the long-run average variable cost curve (LRAVC) for total operating costs is U shaped, as is the fixed investment curve, with costs minimized for firm sizes of 1,600 MW and 3,100 MW respectively. These results are in close agreement with the respective 2,000-MW and 3,000-MW firm size predictions of section I.

It should be noted that scale effects are generally regarded as an important source of cost reductions in transmission. One explanation for the failure to confirm this expectation is that utilities can reduce generating costs by increasing transmission costs.[12] As long as generating costs fall more than transmission costs rise, this trade-off is desirable.[13] One would expect, however, that the effects of this trade-off would appear in the short-run average cost (SRAC) curve but not in the LRAVC curve. Another explanation for the upward-sloping LRAVC curve in transmission is that generating capacity is a poor measure of transmission capacity. While this may be true in general, the limitations placed on the sample in this study should have mitigated this problem.

Clearly the above findings question the natural monopoly status of the electric utility industry and raise serious issues about the appropriateness of current public policies toward it and the generating sector in particular. In addition, one author (Hammond 1972) has suggested that a scale bias exists in utility and government R & D funding and new innovation preferences. From a long-run point of view, it is imperative that the structure of the industry encourage maximum responsiveness to changes in scale, technology, plant-siting needs, fuel avail-

TABLE 1. Summary of Scale Results for the Long-Run Average Variable Cost Curves[a]

Category	Capacity Variables Statistically Significant	Shape of LRAVC Curve	Cost Differential in Mills/kWh by Firm Size				Maximum Size of Scale Effect
Production	Yes	U-shaped	100MW +2.4	1,000MW +0.1	1,600MW Minimum	9,000MW +1.0	2.4
Transmission	No	Upward sloping	100MW Min.	1,000MW +0.05		9,000MW +0.1	0.1
Distribution	Yes	U-shaped	100MW +0.9	1,000MW +0.1	2,600MW Min.	9,000MW +0.1	0.9
Administrative and General	Yes	U-shaped	100MW +0.5	1,000MW +0.4	2,500MW Min.	9,000MW +0.06	0.5
Customer Accounts Expense	Yes	U-shaped	100MW +0.2	1,000MW +0.01	1,700MW Min.	9,000MW +0.05	0.2
Sales Expense	No	L-shaped	100MW +0.1	1,000MW +0.05		9,000MW Min.	0.1
Estimated Total Operating Expense	—	U-shaped	100MW +3.9	1,000MW +0.2	1,600MW Min.	9,000MW +1.1	3.9

Adjusted Fixed Investment per Unit of Capacity

Category	Capacity Variables Statistically Significant	Shape of Fixed Investment per Unit of Capacity Curve	Cost Differential per Unit of Capacity by Firm Size				Maximum Size of Scale Effect
Production	Yes	U-shaped	100MW $111/kW	1,000MW $13/kW	3,100MW Min.	9,000MW $8/kW	$111/kW
Transmission	—	—					—
Distribution	—	—					—

a. The results summarized are those occurring within the range of observation. Note also that firm size measured in kWh has been converted to MW as per note 11.

ability, and pollution requirements. At the very least, the findings of this study suggest that there may be a far wider range of choices regarding industry structure and public policy options than regulations of all phases of the industry.

I have suggested elsewhere the divorcement of electricity distribution from its production, with distribution companies buying power on long-term contracts from unregulated private producers.[14] The results presented here indicate that such a policy option may be more viable than many have suspected. Certainly there is little support for those advocating the merging of electric utilities into thirty or forty massive companies.[15]

It should also be noted that the scale results of this study are likely to be applicable in the future even when the costs of pollution-control equipment become clarified and available for quantitative analysis. This statement is based on the fact that pollution-control costs, even with today's primitive technologies, do not exceed 20 percent of total unit costs for large new generating plants. Unless the scale economies of pollution-control facilities are markedly different from those of the rest of the plant, it is unlikely that the shape of the long-run average cost curve will be radically altered.

Finally, it will be useful to compare the results of our study with those of Christensen and Greene (C and G) in the one area in which they overlap—generation. Controlling for differences among firms, we found that generating costs were minimized for firm sizes between 1,600 MW and 3,100 MW depending on the mix of fixed and variable costs. While C and G did not allow for many differences among firms, they did not restrict elasticities and concluded that costs were minimized for firm sizes as low as 3,800 MW. These results are consistent with our expectation, stated in the introduction, that C and G's treatment of holding companies should shift the scale curves to the right (and flatten it out).[16] Despite differences in approach, the results of both studies are in remarkably close agreement, suggesting that the scale conclusions reached are valid despite potential specification errors contained in each.

Holding Company Results

Table 2 summarizes the regression results for the holding company variables. The holding company variables were of the correct sign in sixteen of twenty-four cases but were negative and significant in only five of twenty-four cases. In one of the five cases (AEP's [American Electric Power Company] total production costs) the negative sign could reasonably be attributed to factors other than holding company efficiencies. Furthermore, in three of the five cases the holding company coefficients attained significance barely exceeding the 90-percent level. In general, there does not appear to be any strong, consistent evidence that holding company affiliation is associated with substantial cost savings.

In the absence of any strong showing of benefit resulting from holding company affiliation, the economies of scale at the firm level would seem to be controlling in terms of the optimal structure of the industry. Furthermore, there is little evidence to support the treatment of holding companies used by Christensen and Greene and others. Our results indicate that, in general, electric utilities

TABLE 2. Summary of Regression Results for Utilization and Holding Company Variables,[a] Short-Run Average Cost Curves

Category	Utilization Variables Statistically Significant	Shape of SRAVC Curve	Cost Differential in Mills/kWh by Degree of Utilization		Maximum Size of Utilization Effect
Production	No	Downward sloping	30%	83%	0.2
			+0.2	Min.	
Transmission	No	Upward sloping	30%	83%	0.2
			Min.	+0.2	
Distribution	Yes	∩-shaped	30%	54%	83%	0.7
			+0.2	+0.7	Min.	

Holding Company Variables

	Any Holding Co.		AEP[b]		Southern Co.	
Expense Category	Coef.	Signif.	Coef.	Signif.	Coef.	Signif.
Fossil Steam Production	−0.09	No	−0.81	Yes‡	+0.17	No
Total Production	−0.08	No	−2.28	Yes*	+0.14	No
Transmission	−0.05	Yes‡	−0.02	No	−0.02	No
Distribution	−0.14	Yes‡	−0.30	No	−0.26	No
Administrative & General	+0.01	No	−0.33	Yes*	+0.09	No
Customer Accounts	−0.02	No	−0.07	No	−0.10	No
Sales	+0.09	Yes*	+0.04	No	+0.03	No
Fixed Investment/Unit of Production Capacity	−13.32	No	+16.21	No	−32.74	No

a. Note that * and ‡ indicate 99% and 90% significance levels, respectively.
b. American Electric Power

owned by holding companies should not be summed and treated as one entity for analysis of scale economies.

Nontraditional Cost Functions and Their Uses

Section II presented several theoretical justifications for our use of a nontraditional cost function for electric utilities. These justifications included the inappropriateness of quantity of output as a measure of scale; the peaked nature of demand; the need to treat durability of capital explicitly; and the need to include specific firm characteristics in the analysis. An additional, pragmatic reason for interest in nontraditional cost functions is their potential usefulness in measuring utility performance and the growing interest of several regulatory commissions[17] in this possibility.

From an overall point of view, the regression analysis summarized in table 1 was highly successful if one uses percentage of regression coefficients with the sign predicted by economic theory, percentage of regression coefficients that are statistically significant, and R^2 as the basis for judging the single-equation models developed.[18] The capacity variables were statistically significant in six of eight regressions and indicated U-shaped long-run average cost curves in all but two regressions.

The regression results for the utilization variables are summarized in table 2. Utilization of capacity was a significant variable only in the distribution cost regressions, and the short-run average cost curves were, over the range of observation, downward sloping for productive costs, upward sloping for transmission costs, and shaped like an inverse U for distribution costs. The magnitude of the utilization effect over the range of observation (0.30 to 0.83) is smaller than that of the scale effects in table 1. This result is consistent with the results reported for plant-level analysis in Huettner 1974. The shape of the SRAC curve in distribution is the only one we cannot explain based on economic theory. The results do suggest that demand patterns may be too complex and varied to be measured by utilization alone.

As acknowledged in section II, our approach did not explicitly deal with durability of capital but at least avoided most implicit assumptions.

As for specific firm characteristics, the regression results support the view that they can be incorporated into the analysis and are important both in terms of statistical significance and magnitude of effect. This should be of interest to those advocating increased use of econometric techniques to measure the performance of utility managements. Yet other results suggest that further problems lie ahead.

For example, one can pretend that the objective of our analysis was to evaluate the management performance of AEP and the Southern Company. Based on the statistical results summarized in table 2, one might conclude that the Southern Company management performance was average since the hypothesis that it was average could not be rejected and even the signs of the coefficients of the Southern Company variable were mixed. Yet the magnitude of the effect of the -32.74 coefficient in the fixed investment regression is not easily ignored. This effect is sizable given the capital intensity of this industry, yet the coefficient is not statistically significant. How should the regulators proceed in this case?

Turning next to the AEP episode, the coefficients indicate lower costs in six of the eight cases and in three border on or exceed statistical significance. Can we conclude that AEP management performance tends, in some cases, to be above average (allowing, of course, for the importance of each cost category in total costs)? Again, how do we allow for the +16.21 coefficient for fixed investment even though it is not statistically significant?

AEP and the Southern Company are reputed to be well managed, yet our regressions have some difficulty confirming this.[19] Perhaps their reputations are not based on fact. Indeed, it might be useful to select some utilities with poor reputations to see if that judgment can be confirmed.

Several basic problems with statistical evaluations of management performance are apparent, however. First, some aspects of performance (i.e., personnel policy, financial results) may not be easily measured or entirely under management control. Second, if high levels of significance are used, we know that few firms will appear in the tails of the distribution, leaving most firms to be defined as average. Use of lower levels of statistical significance would reduce the number of "average" firms but would increase the number of type II errors. Evaluation of performance is always difficult, and management performance reviews by regulatory commissions may be more difficult than expected.

NOTES

1. This chapter summarizes the results of a study published by the author and J. H. Landon. See D. A. Huettner and J. H. Landon, "Electric Utilities: Scale Economies and Diseconomies," *Southern Economic Journal* 44 (1972):883–912.

2. Because a listing of these studies and a review of their assumptions and conclusions have been treated in Huettner 1974 and 1973, the details have been omitted here.

3. The most serious shortcomings of previous economic studies of the electric power industry include scale biases in the treatment of capital costs due to the assumption of equal economic lives for all plants; violation of the principle of cost minimization; and samples concentrating exclusively on plant sizes below 400 MW. For further details see Huettner 1974 and 1973.

4. The degree of integration of holding companies varies considerably. For example, the American Electric Power (AEP) system is highly integrated, the Southern Company system is loosely integrated, and the Central and Southwest System operates with minimum integration of its four operating companies.

5. For a discussion of the myriad forms of interconnection, see Nelson 1969. Also see Huettner 1975.

6. The details of this analysis have been omitted but can be obtained from the authors.

7. Since most firms are integrated to some degree in pools or systems, the unit costs observed at the firm level may be lower than those that would be observed if no integration were present. This factor could alter the shape of the LRAC curve, particularly for smaller firms, but it cannot explain the diseconomies of large firms described later in this section.

8. Of the five empirically based generating-plant studies reviewed in Huettner 1974 and 1973, only two were based on post-1955 data, the most recent including some generating plants constructed in 1965.

9. We are currently collecting data and demographic factors by utility service area.

10. This adjustment consists of the use of an age-weighted Handy-Whitman construction cost index devised by the authors.

11. Two characteristics that could not be included were the number of generating plants and the average plant capacity. Because both are highly (r > 0.8) correlated with firm generating capacity, they were excluded from the analysis. Two holding companies, American Electric Power and the Southern Company, were specifically included in the analysis because of their reputations as efficient producers. Utilization of capacity ranged from 0.3–0.83 for the seventy-four firms, and the range of firm size in the sample varied from 100 MW to slightly over 9,000 MW.

Firm size measured in thousands of MWh-sales has been converted to a capacity measure (MW) by assuming a 60-percent rate of utilization of capacity—the national average—and using the following relationship:

$$\text{firm capacity (MW)} = \text{firm output (MWh)}/0.60 \times 8{,}760$$

This conversion has been used in the discussion of the regression results of tables 1 and 2. Note that there are 8,760 hours in a year.

12. Note that such trade-offs do not exist between distribution costs and transmission or generating costs.

13. Because transmission costs are frequently ignored in discussions of merit loading of generating plants, it is possible that generating costs are minimized and transmission costs simply allowed to fall where they will. Also, since transmission costs are generally much lower than generating costs, any merit-loading practice that minimized their sum would probably result in increased transmission costs and reduced generating costs being optimal.

14. For details of this proposal see Landon and Huettner 1976. Also see Weiss 1975.

15. For example, see Breyer and MacAvoy 1973.

16. Differences in the treatment of holding companies are particularly important since we found that the costs of firms owned by holding companies were not significantly different from those of non-holding company firms. Note also that AEP and some other large utilities have frequently used supercritical, cross-compound, double-reheat generating units. As shown in Huettner 1974, these generating units cost more per kW to construct but unfortunately have not lowered operating costs as expected. Therefore, the types of generating units selected by large utilities as well as diseconomies of scale may account for the results of tables 1 and 2.

17. The Michigan, New York, and Wisconsin regulatory commissions are all investigating various methods of measuring or comparing performance by utility management.

18. More than 70 percent of the regression coefficients were of the expected sign in seven of the eight regressions; more than 47 percent of the regression coefficients were statistically significant in six of the eight regressions, and R^2 exceeded 50 percent in six of the eight regressions.

19. Our regressions have two potential shortcomings, however, for evaluating management performance. The first is that additional utility characteristics, particularly demographic characteristics, should be considered. This shortcoming will be addressed in future regressions when a demand equation is estimated. A second factor to be considered in future work is the utility's participation in power pooling. Since larger utilities have easier access to power pools, the regressions in this study may be slightly biased in favor of larger utilities, that is, the higher costs observed for small utilities may be due to smaller scale as well as inability to join a power pool.

REFERENCES

Barzel, Y. 1974. "The Production Function and Technical Change in the Steam Power Industry." *Journal of Political Economy* 72:133–50.

Boyes, W. J. 1976. "An Empirical Examination of the Averch-Johnson Effect." *Economic Inquiry* 14:25–35.

Breyer, S., and MacAvoy, Paul. 1973. "The Federal Power Commission and the Coordination Problem in the Electrical Power Industry." *S. California Law Review* 46:661–712.

Christensen, L. R., and Greene, W. H. 1976. "Economies of Scale in U.S. Electric Power Generation." *Journal of Political Economy* 84:655–76.

Cowing, T. G. 1974. "Technical Change and Scale Economies in an Engineering Production Function: The Case of Steam Electric Power." *Journal of Industrial Economics* 13:135–52.

Dhrymes, P. J., and Kruz, M. 1964. "Technology and Scale in Electricity Generation." *Econometrica* 32:287–315.

Ewald, E., and Angland, D. W. 1964. "Regional Integration of Electric Power Systems." *IEEE Spectrum*, April, pp. 96–102.

Federal Power Commission. 1966. *Underground Power Transmission*. Washington, D.C.: U.S. Government Printing Office.

――――. 1968 and 1971. *Statistics of Privately Owned Electric Utilities in the U.S.* Washington, D.C.: U.S. Government Printing Office.

――――. 1969. *The Distribution of Electric Power*. Washington, D.C.: U.S. Government Printing Office.

――――. 1971. *The Transmission of Electric Power*. Washington, D.C.: U.S. Government Printing Office.

Galatin, M. 1968. *Economies of Scale and Technological Change in Thermal Power Generation*. Amsterdam: North-Holland.

Hammond, A. K. 1972. "Solar Energy: The Largest Resource." *Science* 178:732–33.

Huettner, D. A. 1973. "Shifts of Long Run Average Cost Curves: Theoretical and Managerial Implications." *Omega the International Journal of Management Science* 1:421–50.

――――. 1974. *Plant Size, Technological Change and Investment Requirements*. New York: Praeger.

――――. 1975. "Scale, Cost and Environmental Pressures: Nuclear and Fossil Steam Plants." In *Technological Change: Economics, Management and Environment*, ed. Bela Gold, pp. 65–106. Oxford: Pergamon.

Johnston, J. 1960. *Statistical Cost Functions*. New York: McGraw-Hill.

Kirchmayer, L. K., et al. 1955. "An Investigation of the Economic Size of Steam Electric Generating Units." *AIEE Transactions* 74:600–09.

Komiya, R. 1962. "Technological Progress and the Production Function in the United States Steam Power Industry." *Review of Economics and Statistics* 44:156–66.

Landon, J. H. 1975. "Changing Technology and Optimal Industrial Structure." In *Technological Change: Economics, Management and Environment*, ed. Bela Gold, pp. 107–28. Oxford: Pergamon.

Landon, J. H., and Huettner, D. A. 1976. "Restructuring the Electric Utility Industry." In *Electric Power Reform*, ed. W. H. Shaker and W. Steffy, pp. 217–19. Ann Arbor: The Institute of Science and Technology, University of Michigan.

Ling, S. 1964. *Economics of Scale in the Steam-Electric Power Generating Industry*. Amsterdam: North-Holland.

Lomax, K. S. 1962. "Cost Curves for Electricity Generation." *Economica* 193–97.

McNulty, J. 1956. "Administrative Costs and Scale of Operation in the U.S. Electric Power Industry." *Journal of Industrial Economics* 30–43.

Nelson, W. S. 1969. *Mid-Continent Area Power Planners*. East Lansing: Michigan State University Public Utilities Studies.

Nerlove, M. 1968. "Returns to Scale in Electricity Supply." In *Readings in Economic Statistics and Econometrics*, ed. Arnold Zellner, pp. 409–39. Boston: Little, Brown.

Olson, C. E. 1970. *Cost Considerations for Efficient Electricity Supply*. East Lansing: Michigan State University Public Utilities Studies.

Weiss, L. W. 1975. "Antitrust in the Electric Power Industry." In *Promoting Competition in Regulated Industries*, ed. Almarin Phillips. Washington, D.C.: Brookings Institution.

Whitman, Requardt and Associates. 1972. *Handy-Whitman Index of Public Utility Construction Costs*. Baltimore: Whitman, Requardt and Associates.

JAMES W. BENSON

CHAPTER 12 | *Energy: Jobs and Values*

Benson examines the impact of alternative energy technologies on the economy and specifically on unemployment. The major part of this chapter summarizes the results of a three-year study that he directed for the Council on Economic Priorities (CEP). The study compared the employment and other economic implications of nuclear and alternative technologies.

To place his summary of the Jobs and Energy study in perspective, Benson then describes the two major "energy constituencies" he sees competing for control of the decision-making processes. He gives a personal view of the politics and values of those resisting more decentralized energy systems, as well as of those who agree with the council's findings.

The first detailed analysis of the impact of electric utility decisions on employment and the economy, the CEP study addressed the question, Should a utility invest in more large power-generation facilities, or does it make more economic sense to invest in "demand-side" energy conservation and solar activities to reduce the need for new capacity?

> The energy question is extremely important, but unemployment is so critical and so painful that we should relate these two issues together in a way that developing the best possible energy policy would also create the maximum number of jobs.
>
> —Senator George McGovern
> before the Senate Joint Economic
> Committee, March 1978

Since World War II, when our nation's energy production started to become more and more centralized, the effects of energy decisions on the economy have greatly increased. Today, with the energy industry accounting for 27 percent of the United States Gross National Product (GNP), its impact on employment and inflation—both in this country and abroad—is massive.

Only recently have we begun to consider the relationship between jobs and energy, but we continue to put our energy money where the corporations—not necessarily the jobs—are. Arguments that conservation and "soft energy" sources—solar, geothermal, etc.—produce more jobs than the harder technologies of nuclear, coal, oil, and the like have been made by advocates of alternative sources, but unfortunately until recently they have furnished little concrete evidence that this is true.

The few available recent studies on jobs and energy typically compared only the direct on-site construction jobs that would be needed to build various energy facilities. It has always been assumed that more workers are needed to build a centralized, capital-intensive energy facility like a nuclear power plant than a decentralized, community-based solar technology. But what about the labor needed to produce each component of a particular energy facility—be it copper tubing for solar collectors or control rods for nuclear power plants? And how many jobs are supported locally with the purchasing power of dollars saved through conservation rather than spent to heat homes?

These and similar questions had not been answered—until now. And the answers are provided by a three-year study by the Manhattan-based Council on Economic Priorities (CEP).

THE *JOBS AND ENERGY* STUDY

The CEP study provides the first comprehensive analysis of the economic and employment ramifications of constructing nuclear power plants and undertaking a variety of energy conservation measures (with some solar energy options). Not only did it examine the labor needed to construct the facilities, it analyzed operation and maintenance jobs as well.

The study focused on the energy choices available to Suffolk and Nassau counties in Long Island, New York. This area is representative of the nation as a whole in that, at the time of the study, crucial decisions were being made about how to meet energy requirements over the next two decades. The key decision faced by Long Island was whether to build two 1,150-MW nuclear reactors, as proposed by the local utility. The potential of meeting Long Island's future energy needs with solar and conservation measures was also being considered.

Clearly the choices faced by Long Island are not unique; throughout the country, communities are now or soon will be choosing from a variety of energy options. Primarily, they are limiting their vision to centralized electricity production with nuclear, coal, oil, and gas, although more localities are becoming aware that they can satisfy much of their energy appetite with conservation and renewable energy resources. Citizens and local officials are struggling to answer the same questions: Which is the least expensive option? Which will have the best impact on the local economy? How many jobs will be created in the community?

To answer these questions, the CEP study analyzed the economic pros and cons of nuclear power, energy conservation, and solar energy measures in the two-county Long Island region. To keep the scope of the study manageable, nuclear power was the only "hard" technology extensively examined since it is the main option immediately faced by Long Island citizens. Gas and oil were considered as part of the present "base scenario" and to provide a point of reference for comparing the impact of conservation and renewable energy measures in the area.

Several scenarios for meeting Long Island's future residential energy demand were compared. The "nuclear scenario" dealt with the construction, operation, and maintenance of the Jamesport twin nuclear power plants. Since the reactors are to take eight years to construct and are to operate for thirty years, the study evaluated all the scenarios over a thirty-eight-year time frame. The main alternative considered was the "conservation scenario," in which approximately thirty-four conservation (improved-efficiency) and solar measures were examined. There were several variations of the conservation scenario as well: the "conservation/electric scenario" analyzed only those measures that specifically saved electricity, thus allowing a direct comparison of conservation and continued consumption of electricity; the "solar scenario" investigated only solar measures, such as space and water heating; the "CEP/NY scenario" examined only those conservation measures (primarily weatherization) entitled under the New York State Energy Conservation Act to be financed by the local utility company.

As noted above, the employment figures are not limited to construction but are the composite of three broad employment categories examined in the study. Construction jobs were covered in the first category, "on-site" employment, which consisted of the jobs required to build a power plant or to install conservation or solar measures. The second, "direct" employment, analyzed the jobs required to produce the materials and to manufacture components for each energy technology. The third category considered was "induced" (indirect) jobs, which are created and supported by the groceries, gasoline, clothes, utility bills, etc., purchased by workers with their wages from constructing a power plant or installing conservation or solar measures.

Before the direct and indirect employment could be calculated, CEP had to determine both the regional and national economic implications of each energy choice. It was therefore necessary to itemize the "whole-system" cost of each energy strategy by adding the costs of materials, labor, overhead, and profit. The dollar value of energy-saving measures also was determined in order to calculate their impact on household discretionary income, or income after taxes and savings. Reduced energy expenses translate into more money available for the household budget, which is used to purchase other necessities, thereby creating more indirect jobs in the local economy.

To calculate the costs associated with a comprehensive energy-conservation and solar-energy program for the residential sector,[1] CEP assumed that each home has installed up to twenty conservation measures, twelve improved-efficiency appliances, and two solar measures (depending on how many measures have already been installed). The study considered only proven measures that are widely available today. (Technologies yet to be perfected, such as photovoltaics and bioconversion, were not included.)

The study divided residential units into four categories: existing single-family, existing multi-family, new single-family, and new multi-family. The current energy use characteristics for each of these categories was determined based on an estimate of how many homes already had installed each conservation measure. Assumptions were also made about the number of conservation measures each household unit needed in order to be tightly sealed and insulated. All major appliances were to be replaced with more efficient ones as they wore out; all heating, ventilation, and air-conditioning systems were to be modified and maintained for optimum efficiency; residents were to reduce their use of energy for lighting; and solar energy was to replace some conventional water and space heating. New homes were to be constructed in accordance with stringent conservation specifications.

Wages (union scale) were set for each category of labor, and the type and amount of labor needed to install each of the conservation and solar measures were estimated. The need for and cost of materials like fiberglass, pumps, and lumber were determined, and the percentage of each material that could be produced locally was calculated. A complex series of computer programs was devised to perform these calculations for the eight hundred thousand homes in the Nassau/Suffolk region. One program calculated a detailed bill of goods for materials and labor to install the various measures; another used these results to compute regional economic multiplier effects (the effects of money flowing through the economy, being spent and respent to stimulate more economic activity and to support a variety of jobs). Other programs dealt with energy savings for gas, oil, and electricity and with national economic impacts.

CEP used these programs to determine the year-by-year material requirements and costs, labor requirements, energy savings, and regional and national impacts of all the scenarios over thirty-eight years. The resulting figures are not predictions or forecasts; they were created from a reasonable set of assumptions in order to provide a basis for determining the jobs created and the energy saved per dollar of investment, both locally and nationally. The total cost of the conservation scenario is $4.04 billion.

A similar process was followed for nuclear power plants. Construction, operation, and maintenance costs were obtained from the Long Island Lighting Company, which proposed the plants, as well as from various studies. These were then divided into other categories, such as materials, labor, and taxes. Annual on-site labor requirements were determined, and material requirements were analyzed to estimate those that could be produced in the local economy. A detailed bill of materials was prepared for both the regional and national economic models, and the year-by-year costs and labor requirements were calculated over a thirty-eight-year period (eight to construct, plus thirty to operate). The total cost of the nuclear scenario is $4.01 billion.

The results of the study were surprising even to staunch proponents of solar energy and conservation. The conservation scenario was found to stimulate 62 percent more economic activity and 50 percent more employment in the national economy than nuclear power, per dollar invested. The study also showed that simply substituting conservation measures for an energy mix of oil, gas, and electricity, where appropriate, would provide *two and one half times* more jobs nationwide than the conventional fuel mix.

Table 1 shows the various amounts of employment supported by different types of expenditures considered in the CEP study. The largest amount of employment is supported by household spending (personal consumption expenditures). The smallest amount of employment is supported with purchases of fuel oil. In the New England area, where most oil is imported, only a fraction of the employment is retained in the national economy—the rest is exported in the form of cash payments for the imported oil.

Table 1 can be used to obtain a rough estimate of the employment impact of changing expenditure patterns. For example, in 1978 utility companies in the United States requested a total of $4.6 billion worth of rate increases. If these were granted, household expenditures would be shifted from personal consumption expenditures to the electricity category. This would result in the loss of 231,380 jobs from personal consumption expenditures and the addition of 110,400 jobs in the electric utility sector, for a net loss of 120,980 throughout the United States economy.

Table 2 shows the effect of two types of shifts in expenditures. The first shift, from personal consumption expenditures to a mix of energy consumption, results in an overall loss of 34 jobs per million dollars per year. This means that for each consumer price increase of $1 billion for oil, gas, and electricity, 34,000 jobs are permanently lost to the national economy. The second column of table 2 shows the effect of shifting expenditures away from mixed energy consumption into a comprehensive conservation and solar program. For each $1 billion shifted from consumption to conservation, 31,000 jobs are created and maintained.

The cost of the conservation scenario, over thirty-eight years, would be ap-

TABLE 1. Employment per $1 Million Dollars Spent

Type of Expenditure	Labor Years
Personal Consumption Expenditures	50.3
Conservation/Solar Scenario	48.8
Nuclear Power Scenario	31.0
Electric Utility	24.0
Mixed Energy Consumption[a]	17.7
Natural Gas	15.5
Fuel Oil	14.1

a. Oil (49.6%), Electricity (39.3%), Natural Gas (11.1%).

TABLE 2. Employment Changes due to Expenditure Shifts (Labor Years per $1 Million)

Economic Sector	From Personal Consumption Expenditure to Mixed Energy Consumption	From Mixed Energy Consumption to Conservation
Agriculture	− 2.3	.2
Mining	1.0	− .9
Construction	.2	18.1
Manufacturing	− 9.4	15.6
Transportation	− 1.0	.1
Communication	− .8	.1
Public Utilities	4.6	− 5.0
Wholesale & Retail	− 13.3	1.4
Personal & Professional Services	− 12.3	3.2
Government Enterprises	1.1	− 1.8
Total	− 34.2	31.1

proximately $5 billion (in 1979 dollars). The total potential for energy savings would be 111 billion kWh, 532 million MCF of natural gas, and 17.3 billion gallons of fuel oil, or over $24 billion ($.065/kWh, $4.00/MCF, $.85/gallon).

Nuclear power is widely perceived as a means of greatly reducing our dependence on foreign oil supplies, even though only 10 percent of the United States oil supply is used to generate electricity. In the Northeast, however, nuclear power could significantly reduce the need for imported crude oil, because most of its electricity is oil-generated. Consequently, CEP compared the potential impact on oil imports of nuclear power with a conservation/solar strategy and found that the conservation scenario, which includes *residential* opportunities alone, could reduce foreign oil imports twice as much as the proposed Jamesport plant.

Further opportunities for energy conservation and solar energy use in commercial and industrial establishments could reduce the need for imported oil to a far greater extent than *both* units of the proposed Jamesport plant. Table 3 indicates that costs per unit oil displaced by the residential conservation and solar measures are 42 percent of CEP's estimate for such costs for the Jamesport plant.

In this analysis nuclear power is substituted for oil-fired electric generation. If we were to attempt to substitute nuclear for direct uses of oil, such as in transportation and use for home furnaces (as is necessary to make a significant impact on the nation's oil-consumption level), it would cost even more per unit of oil displaced. This is a consequence of the great inefficiency of generating high-quality electrical energy, only to reduce it back to low-quality heat energy.

In the United States economy as a whole, a strategy such as the one aimed at improving the efficiency of energy use in Nassau and Suffolk counties would create one and a half times as much employment per barrel of oil displaced as would substitution for oil with nuclear electricity supply. Local employment

ENERGY: JOBS AND VALUES

TABLE 3. Oil Displacement

Energy Scenario	Total Cost (Millions 1976 $)	Btu's Displaced	Barrels of Oil Displaced	Present-Worth Cost per Barrel Displaced
Jamesport 1 & 2	$5,297	3.61×10^{15}	581×10^6	$16
Conservation and Solar	$3,172	4.19×10^{15}	708×10^6	$ 6.80

created within Nassau and Suffolk counties would be more than twice as much per barrel of oil saved.

Figure 1 shows the amount of on-site labor required for the conservation scenario and the nuclear scenario. Total on-site labor for conservation would be 75,510 labor years compared with 13,240 for nuclear. As we can see, not only is more labor required during the installation of the conservation scenario, but more labor would also be required during the operation and maintenance period.

The most important difference between conservation and new supply programs is that conservation pays itself back, while new supply simply costs more permanently. For example, the cost of the conservation scenario is recovered through energy savings in approximately eight years—less time than it takes to build new nuclear or synthetic fuel plants.

For planning purposes it is essential to consider shifts in expenditures and

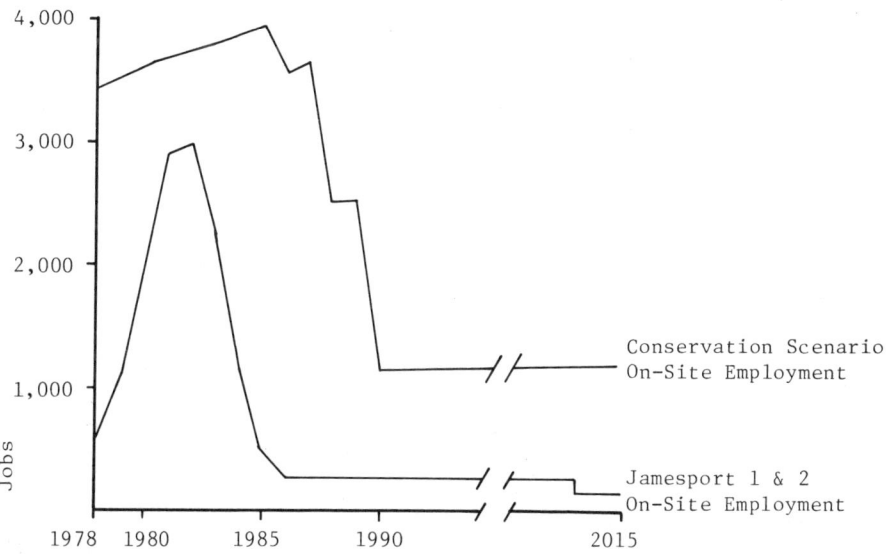

FIGURE 1.

the flow of money over the time period under consideration. First, money must be shifted from its present use into the construction of a plant or the installation of conservation measures. Next, the money is spent for one of the scenarios.

The CEP analysis found that more employment is created through the installation of conservation measures than through the construction of nuclear power plants. Additionally, more of the materials are supplied locally. In other words, there is more of an economic drain on the local economy with the nuclear scenario because more money flows out of the local economy.

As the scenarios are being implemented, another benefit to the national economy is in the manufacturing sector. The CEP analysis found that more employment is supported, per dollar of investment, with the conservation scenario.

A significant difference now becomes apparent. Immediately upon installation of each conservation measure, energy savings begin to accrue. This begins to pay back the investment while reducing imports, if the savings are in oil or electricity generated from imported oil. However, eight to twelve years from the start of construction the new plant begins to produce. Another significant difference becomes apparent. The output of the plant must be purchased. This in turn causes a further decrease in household discretionary expenditures along with a resulting decrease in employment, and an increase in the flow of money out of the local economy.

As fuel-oil prices rise, the proportion of money that goes toward wages and profit will decrease because the same amount of labor is required to supply the oil. Similarly, as the cost of building and fueling nuclear power plants escalates, the money spent for nuclear power will create even fewer jobs per dollar invested. This means that constructing nuclear reactors would perpetuate the outflow of money from the local economy, creating fewer jobs in the long run.

The average wages for constructing nuclear power plants are higher than those associated with the conservation scenarios, but the conservation scenarios are significantly more labor intensive. Therefore a higher proportion of the total cost goes to wages. In addition, a smaller proportion of the materials needed for nuclear plants is available locally—and *all* money used to purchase nuclear fuel leaves the region.

The CEP study conclusively shows that switching to conservation and solar energy is cheaper, creates more employment, and is far more beneficial to the regional and national economies than is continued consumption of the present fuel mix or the construction and operation of nuclear power plants.

Additionally, conservation and the use of renewable energy can immediately begin to reduce our dependence on imports. Conservation and renewables are clearly safe and more environmentally benign than nuclear, coal, and synthetics. Finally, they lend themselves to decentralization, meaning there is less need for multinational corporation and federal government involvement and more opportunity for citizen involvement in decision making.

With all these factors favoring conservation and renewables, it seems surprising that since the formation of OPEC there has not been a massive effort to move in this direction. It now appears that all the necessary information is available for creating a national policy with specific programs to implement it.

TWO PATHS TO THE FUTURE

Unfortunately, translating the results of the CEP study into national policy will be an extremely difficult task. Our society has developed numerous barriers to a solar/conservation transition, of which putting our energy destiny into the laps of corporate decision makers is one example. But the most serious barrier is the way we think about energy. Each of us, because of our own personal experiences and our selective exposure to the facts, has different perceptions not only of what the energy options are but also of what the overall goals of society are. Our individual prejudices shape our view of the world so greatly that we tend to argue in clichés when we debate issues that affect us. This is particularly true when it comes to energy, which is inseparable from the most pressing sociopolitical issues of the day. Adversaries appear to be debating the technical pros and cons of different energy options while actually trying to resolve fundamentally incompatible perceptions of reality.

Two major doctrines on energy issues seem to prevail today. The views of those who subscribe to either of these notions are sacred and cannot be compromised, thus making a solution to our energy dilemma most difficult to find. We can generally agree that energy is a means and not an end, but there the agreement stops. Proponents of both perceptions have their own answers to the real question: Energy to what end?

Corporate executives and, not surprisingly, the federal government generally subscribe to the notion that the energy problem is essentially a problem of developing new supplies to meet the ever-expanding energy demands of a growth-oriented consumer society; simultaneously, conservation technology needs to be adopted, when economical. This view of the energy picture leaves little room for the changes implied by the CEP study. It entails the following broad assumptions:

- The benefits of a high-consumption, high-technology society are obvious and should be generally agreed upon.
- Society is a "zero sum game." For any individual or institution to gain, a competitor must lose. Short-term interests are maximized, long-term consequences minimized.
- Major decisions should continue to be greatly influenced by the financial and military-industrial sectors, because a healthy industrial sector means that benefits will trickle down to the less fortunate.
- The high-consumption, high-growth society provides the only hope for raising the poor to a higher state of material, and therefore social, well-being.
- A technological approach to the problems of ensuring adequate energy supply should receive strong public support because of the proven accomplishments of technological solutions and larger institutions in the past.
- The inevitable environmental costs of large energy projects must be chosen

over the economic and social penalties of running out of energy, and regional sacrifice (e.g., Alaska and the Western states) must be accepted to ensure continued GNP growth.
- The goals of ongoing economic, material, and technological growth provide an adequate incentive to resolve land-use and water-rights conflicts and other related problems to industry's benefit.

When the progress of a nation is measured by criteria such as GNP, industrial productivity, and energy consumption, the decision makers tend to view any questioning of these measures as an assault upon the system itself. The civil rights movement, the anti-Vietnam protests, the exposure of political and corporate corruption, and the resistance to centralized energy systems such as nuclear power are perceived as a threat to the status quo—a challenge to their very legitimacy—as the following excerpt from a report of the Trilateral Commission (*The Crisis of Democracy*, 1975) reveals:

> A significant challenge comes from the intellectuals and related groups who assert their disgust with the corruption, materialism and inefficiency of democracy.... The advanced industrial societies have spawned a stratum of value-oriented intellectuals who often devote themselves to the derogation of leadership, the challenging of authority, and the unmasking and delegitimation of established institutions.... This development constitutes a challenge to democratic government which is potentially at least as serious as those posed in the past by the aristocratic cliques, fascist movements and communist parties.... The governability of democracy seems dependent upon the sustained expansion of the economy. Political democracy requires economic growth.

The study even goes so far as to suggest that continued economic growth may warrant the restriction of some of our most fundamental rights:

> The vulnerability of democratic government in the United States thus comes not primarily from external threats ... but rather from the internal dynamics of democracy itself in a highly educated, mobilized, and participant society. We have come to recognize that there are potentially desirable limits to the indefinite extension of political democracy.

To the extent that the assumptions of the Trilateral Commission are accepted, a drastically lowered energy demand (e.g., future national demand near or lower than the present level) would be both unacceptable and implausible because of the perceived implications of lowered economic productivity and increased unemployment. Instead, it would seem reasonable to expect a continuing rise in GNP and a corresponding growth in energy demand, with a large amount of future energy supply coming from massive coal and shale developments, breeder reactors, and other advanced energy technologies—at something like the levels projected by many U.S. government agencies. It would also be reasonable to foresee considerable promise in large, centralized solar power installations, including such exotic technologies as oribiting satellite-based solar collectors.

Many people are convinced that the growth of the economy is so crucial that they are willing to take significant military, environmental, public health, and safety risks to protect it. However, more and more people are questioning the value of breeder reactors, massive shale and synthetic-fuel development, exotic solar technologies, and the military-industrial control of our resources. Objections to large-scale fossil fuel and nuclear projects on social, environmental, and health grounds are becoming commonplace as a different perception begins to emerge. This view of the energy dilemma holds that the energy problem is basically a matter of increasingly intolerable social, environmental, and economic costs of continued energy and material consumption. Energy demands need to be cut, with changes made equitable by supporting legislation.

Increasingly, people are becoming convinced that the energy problem is but one manifestation of a far more fundamental crisis involving nothing less than the basic assumptions and goals of all industrialized societies. Major changes in our values and institutions, they believe, must be made soon if we are to create a more democratic and ecologically sustainable society and avoid global disaster.

People of this persuasion have a strong awareness of the "new scarcity"—of physical resources and of the capacity of the environment to tolerate waste—and believe that these problems are qualitatively different from the scarcity problems "solved" by modern industry. They would enthusiastically put the policies implied by the *Jobs and Energy* study into practice, but they are politically unorganized. They are a threat to the system because they believe:

- The further centralization and electrification of the national energy system is environmentally and socially destructive and decreases national security.
- Future energy needs (as opposed to artificially stimulated wants) should be provided by a decentralized energy system characterized by renewable energy utilization and democratic control.
- The public interest is not being served by elected representatives who owe their support to vested interests.
- There should be more equitable distribution of the earth's resources among all nations, including Third World countries.
- Materialism should give way to a simpler, more wholesome life-style characterized by appropriate technology.
- Life is a "non-zero sum game," that is, everyone can win in the quest for energy independence only if the well-being of the global ecological system is maintained.
- Meaning and commitment are lacking in modern industrial society.
- A strong "ecological ethic"—an identification with nature, fellow beings, and future generations—is the only way to ensure the survival of society.
- People are "caught in the system"—they are impotent and are being victimized by large, impersonal institutions.

- Capital-intensive "big" technologies—from assembly lines to centralized computers—are dehumanizing and impoverishing.

Out of the sense of alienation and impotence caused by modern industrialized society comes a desire to own and/or control the technology and institutions affecting one's own life. Thus, whether energy systems are controlled by the community or individually, or are owned and controlled by a large remote private power company, becomes a crucial issue.

The symbolic importance of the decentralization issue should not be underestimated. Solar energy is democratic: it falls on the rich and the poor, the weak and the powerful. Decentralized solar energy symbolizes keeping it that way—maintaining independence from the "big system." Just as the principle of control and ownership of property was essential to liberty when our country was founded, so the principle of control over energy supply is a precondition of liberty for those who hold this perception.

CHOOSING THE PATH

Each of these two perceptions has to be respected, in some sense, since each *appears* to "fit" the model of the world shaped by the person holding that view. Consequently, no solution to our energy problems will be "correct" to everyone. At the same time, however, it is imperative that some reasonable criteria be applied to evaluate our different energy options, for each one leads society to a drastically different future. At least two fundamental questions must be answered as we plot the course our energy future will take:

1. *Does the choice in the long term lead toward societal and ecosystem adaptability, and hence toward survival?*

The major shortcoming of our energy decision makers to date is that they have failed to look at the long-term consequences of their actions. Nuclear power originally was viewed (and to some still is) as the panacea for our ever-growing energy "needs." But the possibility of "Class 9" accidents, and of never finding that elusive repository for radioactive wastes, was seldom considered. One argument often put forth in favor of the continued-growth view is that, like the skater on thin ice, we cannot risk stopping. But we must be able to adapt to changes in our environment if we are to survive, and the "thin ice" argument is equivalent to admitting that we have *already* lost our ability to adapt. The principle of adaptability is central and cannot be ignored, regardless of an individual's picture of reality. As the current decline in fossil-fuel resources indicates, an energy-intensive industrial society is less able to adapt to changes in its environment than is a more decentralized, loosely structured society.

2. *What are the full social benefits, costs, and risks of a shift toward reduced total energy consumption and a reliance on small-scale, decentralized, renewable energy sources?*

This question cannot be answered without assessing the major goals and priorities of our society, and our basic beliefs about humanity's place on the planet and in the universe. The issue of shifting to solar energy and other decentralized renewable resources does not involve technological and economic factors alone; it involves the fate and the future of Western civilization.

The question of how much energy we as a society should consume is not simply a matter of how much we need to continue our profligate ways; it points to the need for a fundamental re-evaluation of *where* society is going. We have but two energy choices: a nuclear-dominated, high-technology energy future, with solar energy playing a secondary role at best, or a solar-oriented decentralized "soft" future, with reduced energy demand and an emphasis on ecological stability and quality of life. If we accept the premise that we must look at the long-term consequences to future generations of our energy (and all other) decisions, that curbing inflation, equitably distributing resources, ensuring health and safety, and alleviating unemployment are necessary to cure our political and economic ills, then the choice of energy path seems clear.

The number of studies documenting the inevitability and attractiveness of a soft, or decentralized, energy future is large and growing. The answers to fundamental energy questions seem to favor a rapid turn to this future, which can have only positive impacts on inflation, the environment, and unemployment. Rather than "the system being the solution," it is more likely that the systematic and regular re-evaluation of our needs and goals is the only solution, and through the decentralizing of electrical production we are increasing our flexibility and are gaining more understanding and control of our destiny.

NOTE

1. To make the project more manageable, factors were introduced to limit and simplify the research. Whereas energy consumption usually is divided into four sectors (residential, commercial, industrial, and transportation), here it was restricted to an analysis of the residential sector, which consumes 48 percent of all electricity in the Nassau/Suffolk region. Commercial buildings, industrial processes, and transportation energy use patterns were not considered.

E. F. LINDSLEY

CHAPTER 13 | *Planning Practically for a Decentralized Electrical System: How Past Experience Can Guide Us*

In this analysis of the problems and possibilities of a decentralized grid, Lindsley draws on a lifetime of hands-on experience with alternative energy and related industries. The chapter deals primarily with the management problems of the technologies themselves, rather than with grid management issues (as discussed in chapter 5 by Morris and chapter 6 by Sørensen). Lindsley warns that the alternatives to large, centralized power production have their own difficulties. He also presents helpful strategies for making a decentralized system more workable. Of particular importance is his call for a national or regional service industry to maintain these systems.

Studying the history of technology can be valuable as we contemplate the possibilities for decentralized electricity production. We would be foolish not to make a broad survey of our considerable national experience in localized power production and in other fields using randomly placed prime movers. This experience can provide insight into the problems that are likely to be encountered in a revival of decentralized power production.

The pitfalls and mistakes are as important as the breakthroughs in planning. In my experience, one lesson is extremely important; for the successful operation of any type of distributed modular power units, some type of large central control and service agency will be needed. It must be divorced from local political influence and must concentrate on technology.

We should recognize that much of our past energy development was based on small-scale, less sophisticated, and less costly projects than we now envision. Mistakes were made, of course, but were usually limited disasters, painful only to those few individuals directly concerned. An example is the 1,000-kW Putnam wind generator, which for sixteen months during World War II fed power into the lines of the Central Vermont Public Service Corporation. Located atop Grandpa's Knob near Castleton, Vermont, it ultimately failed dramatically because of the breakage of a blade that was known to be defective but could not be replaced because of wartime material shortages. This failure, never well understood by the public, cast a shadow over wind-power development for thirty years. The discouragement of further experiments was more damaging in the long run than the immediate financial impact on the small family-owned company that built the generator.

Today such public misconceptions could be greater in scale and expense, owing to mass media coverage and user interest. Previous generations were more willing to accept new technology as an evolutionary process and expected only step-by-step improvements. Today we somehow expect technical perfection from the first throw of the switch, and when it fails we are bitterly chagrined.

Another example of this problem is the ill-famed Bay Area Rapid Transit (BART) in San Francisco, originally touted as the ultimate in automated public transportation but troubled for years by technical problems. Though many of the problems were solved, the initial image of failure lingers. Both cases emphasize the basic need for highly centralized monitoring of fabrication, procurement, and service planning. The Putnam windmill was a victim of wartime conditions, which made it necessary to scatter fabrication of parts among producers on a catch-as-catch-can basis. Later there was no way to replace vital components or perform routine maintenance, repair, and replacement. The BART problems also demonstrated the hazards of divided design and fabrication responsibilities in a complex system. Even worse was the later division of service and maintenance responsibilities, resulting partly from politics and partly from the specialized technical expertise of diverse suppliers. Since both of those systems were unique, parts had to be specially made.

These observations stem from considerable personal experience in the power and energy field. I have had the opportunity to survey small-scale power production from two distinct viewpoints, first, through many years of field service supervision for a pioneer builder of on-site, total-energy, natural gas- and diesel-powered generating plants, and now as engineering editor of a magazine devoted to reporting new technology including all facets of advanced energy planning.

In my first position the total-energy concept was to deliver raw energy—natural gas or diesel fuel—to the use site and convert it into both electrical power and useful heat, a cogeneration technique that could be nearly 80 percent efficient. Such on-site, total-energy plants were deemed ideal for apartment complexes, shopping centers, sports arenas, schools, and factories, where the heat could be used for processing. However, dismaying gaps developed between the thermodynamic ideal and the real-world experiences with on-site power. The thermodynamics were sound; the troubles all pointed to a need for a strong central control and service agency.

Happily, other experiences in my background seemed to extrapolate in extremely practical ways to decentralized power generation and in fact supported the probability that given the necessary organizational structure it can be made to work and work well. These experiences involved dispersed oil field pumping and agricultural irrigation power. Each of these applications required multiple modular on-site prime movers not unlike what one would expect to find in decentralized power generation. In any case, both the good and bad experiences can provide us with direct historical backgrounds that may, if we're wise enough, apply almost unchanged to widely distributed electrical modules.

Today in my second occupation I enjoy that editorial duty "to view from afar" while remaining aloof from the demanding quotidian technical problems and assessments of responsibilities for them. My present task is to scan the horizon for new energy-producing concepts. It is notable how many energy concepts are best suited—if not, indeed, limited—to decentralized units. Among these are photovoltaics, sun-powered engines, bio-plants, wind power, and waste-heat recovery.

Maintaining a historical perspective turns up some fascinating, if ironic, insights into our wastefulness in the past. For example, the venerable Stirling engine (which pumped drinking water for cattle long before the American farm windmill) now offers, in a very elegant wrapper, one of the more remarkable hopes for converting direct sunlight to AC power. As envisioned and now being tested, the Stirling would certainly constitute a decentralized multiple-module energy source.

My emphasis on these two personal viewpoints is intentional. Imagining new energy sources is exciting and allows the mind to race unhindered by constraints, but applying these alternatives remains constrained by real-life hardware, technical and economic factors, human weaknesses, and the social matrix on which we would impose our dreams. In the often harsh world of mechanical reality, a honeymoon may end abruptly.

If I seem hard-nosed regarding decentralized power generation, let me repeat a statement I made when speaking on that subject at Wesleyan University: "Until you've walked into a totally dark generator room with a flashlight in one hand and a toolbox in the other, you haven't had a firsthand experience with on-site power."

We are often presented with bucolic images of old New England's water-powered mills again turning picturesquely, each producing its share of 60-Hz, 3-phase, voltage-regulated power and feeding it into a wide-flung grid. Just over the hill there would be glistening photovoltaic arrays and perhaps a farm of spinning wind generators all adding their bits and none of them polluting or exhausting nonrenewable resources. In some of these pictures, at least in the literature I have received, happy communal dwellers are gathered about the power plant, strumming guitars, grinding natural foods by hand, tending children, and, presumably, ready to spring to instant remedial action at the first flicker of a needle on a power distribution panel. They will, of course, have the exact part number relay, governor component, or Timken bearing indexed and at hand together with the tools and needed know-how to repair the ailing generator and phase it back onto the bus. Their wages will be the sheer joy of contributing to the overall good of the energy scene.

If this bit of satire (for which I apologize to my friends in the counterculture) seems unfairly drawn, it is not. The truth is that with certain minor modifications that really is the way it used to be in many small, local power plants. While researching a recent article on small hydropower, I was amazed to learn that it was not at all uncommon for such plants to be manned and backed up by family groups. In fact, quite a few such small, private hydro plants operated on contract to the local utility and had a history of feeding a modest amount of power to the grid for many years. Without exception there was a family patriarch who had perceived early on that his pond, stream, and water flow were capable of driving one or two turbines. He was an enthusiast, he was dutiful, and he was reasonably skilled in maintaining simple generators and water turbines. Equally important, he was willing to risk life and limb if need be to fight ice jams, spring floods, storm debris, and lightning to keep the plant operating. He mixed and poured concrete and built rock walls to repair the dam. His pay was, perhaps, about one cent per kWh and it about matched his wife's egg money.

Recent publications have pointed to studies, some by the Army Corps of Engineers, showing literally thousands of sites in the Northeast with sufficient water head for power generation. The potential must be viewed in the light of older small hydro plants if we are to hope for their direct application to decentralized power. In particular, certain common factors in the older plants could preclude any immediate revitalization:

- Many provided only mechanical power to drive a gristmill, sawmill, or workshop.
- Many of those producing electricity produced DC with badly managed and erratic voltages and outputs unsuited to feeding a grid.
- AC generators were primitive, poorly regulated, and poorly governed.
- Many operated during the day but had to shut down at night to allow the millpond to refill.
- Most used generators and switchgear produced in small quantities and were disparate in parts and design. They had few features in common.

I invite those who doubt this assessment to tour their own areas and make their own inspections. I will cite one experience that is typical of what I found. I was offered fifty acres of prime farmland, a house, a dam, a pond, and a complete generator installation with two Leffle turbines for $50,000. The old homestead and power plant were impossible for the current owner to maintain any longer. This was an operable hydro plant that had supplied power to the utility for one cent per kWh for many years, but the man who had devoted his life to it was dead. As the present owner explained, the old man had always been there to look after the turbines because, after all, "He had cows to milk."

Although not tempted, I was curious, and on inspection of the plant I saw a control panel with old-fashioned copper knife switches, a hand-operated field current control, and a pair of light bulbs for phasing. Cobwebs and the dust of fifty years covered everything. Huge, flat, wide leather belts ran from the shafts of the old turbines to the generator. One couldn't view the scene without nostalgia. Of course, the equipment might have been replaced with modern gear but the irreplaceable factor was the devoted old man.

Who would fill the shoes of such men at scattered sites today? And how many such sites would be needed to pay their wages? What we're talking about is economy of scale. The old man was actually on duty and sometimes worked three eight-hour shifts, seven days a week, if we look at it in punch-in and punch-out terms. If we paid three men $30 a day, that would be $90 per day, plus social security, unemployment compensation, health insurance, and retirement benefits; we would also have to cover them with liability insurance in case they erred in handling the dam gates and flooded the town downstream. I cannot estimate how many kWh per day it would take to pay just these costs, but I know that the total capacity of the hydro plant couldn't touch it.

Moreover, this would not cover taxes, return on capital, wear and tear, amortization, reserves for replacement, or emergency repairs. The owner told me that he had spent $15,000 on minor dam repairs but that at least another $50,000 was needed; and that was obvious. In short, this small hydro plant was just too small to support its own existence. Small may be beautiful but in many cases it reverses the economy of scale.

If any historical conclusion is to be drawn from this, it is that for any given area where revival of preexisting plants seems feasible, the initial step must be a thorough engineering and economic survey of the plants. This would undoubtedly rule out some old power sites better suited to nostalgic postcards and calendar art than to power generation. If their output capability cannot support modernization of the plant and subsequent maintenance within a reasonable payout period, they are not viable.

One counter to this argument is that modern technology can reduce the need for manpower by automation. That's the way BART was supposed to run—without a motorman on each train. In the case of small hydro, timers, relays, little microchips that think, automatic sensors, and the like would replace the old man who came running whenever he saw sparks or smelled smoke.

This is possible to some extent, but the sad history of shopping centers, retirement villages, and other total energy installations where total reliance was placed on automation testifies to the real-world necessity for human monitoring. We have only to remember that in the most advanced and sophisticated application of automation—our space program—more missions were salvaged by human presence on board than went smoothly and without intervention. Later, at the Three Mile Island nuclear plant, we witnessed the most disturbing testimony yet to the folly of total dependence on automation and to the importance of strong, central human control. Here the former did not work because of something as common as a stuck valve, and the latter proved ineffective owing to overly divided responsibilities.

There is no reason to assume that things would be different at decentralized power sites. Experience shows that it is seldom the big things that fail. A hose or pipe connection opens up, dust or moisture gets into a relay, a transient voltage excursion knocks out power to a control that fails to reset, or someone who doesn't know the first thing about what he's doing turns a rheostat knob the wrong way or shuts down a vital pump.

If all this sounds like haggling over mere mechanical details and blindness to the overall elegance of decentralized power, I recognize that many proponents of decentralizing make their livelihoods in fields where the obstinacy of mechanical things is not a daily factor. They prefer to look at the big picture. Unfortunately, the big picture is painted by thousands of tiny brush strokes in the world of technology, be it at the bailing wire or the microchip level. I recall flying a technician to Cape Kennedy when one of our moon-bound spacecraft was ready to launch but on hold because of a niggling problem with a voltage regulator on an engine driving a back-up generator. The men in white coats had spent three days trying to adjust a plastic screw that turned out to be only a dust plug concealing the real adjusting screw underneath. Murphy's Law has not been repealed by modern technology. To summarize:

- Regardless of the prime mover—hydro, photovoltaic, wind, whatever—it must produce enough power to support its own maintenance costs. Hobby operators and friends of the cause will never contribute significantly.
- Power sources must be maintained *constantly*, not fixed when they break down. The alternative is doubling up equipment for backup during unexpected down time. The latter is seldom economical.
- Power sources must be maintained only by technically competent, professional personnel.
- The above, requiring a skilled service industry, is practical only with a large economic support base.
- A high level of standardization of equipment is a must.
- Power sources must be sufficiently close together for daily monitoring and maintenance without excessive crew travel.

If we acknowledge these realities, how can looking backward guide us in the future? Probably as good a learning experience as can be found is the above-mentioned proliferation of on-site total-energy plants in the 1950s and 1960s. We can picture three or four engine-driven generators cutting on and off the line automatically in response to load demands. The waste heat from the engine's exhaust, water jacket, and oil cooling was routed through a low-pressure boiler and used for building heating, processing, or absorption air conditioning. Naturally the operation was highly automated with the most sophisticated state-of-the-art devices. Moreover, the sales engineers painted a rosy picture that no serious operating problems would arise and that no one other than the janitor would be needed, and he would be required to look in only occasionally.

Except for the fact that these plants were not feeding into the power line, they closely resembled what might be expected with distributed modular power outfits. One advantage was that the prime movers—heavy-duty engines—were products of long-standing technology, not subject to nature's whims as with hydro or wind and not exotic such as photovoltaic. What then were the historical problems?

Financing these plants required detailed operating and maintenance cost projections (largely guesswork based on ideal operation) for as long as twenty years into the future. Looking back, these projections were unrealistic because of inflation and unforeseen events.

- Regardless of automation, someone had to provide routine periodic inspection, service, and repair.
- Regardless of built-in reliability, personnel with very high level skills, both mechanical and electronic, had to be on call seven days a week to perform emergency repairs. In those areas with highly unionized specialties this often meant at least three or four men.
- For the same reason, whoever had the backup and maintenance responsibilities had to maintain a total and high-cost inventory of critical parts.

Thus, although total energy looked very attractive in theoretical thermal terms, it was actually fraught with real-world problems. Typically the unskilled personnel assigned to monitoring were either too overwhelmed by the complex of valves, switch gear, pumps, and associated devices to do anything, or they were unable to evaluate what they saw. Some individuals, cursed by an incurable case of "tinkeritis," sooner or later introduced mysterious troubles for which the technicians couldn't find a source because they couldn't imagine anyone "doing that." Does this sound familiar after Three Mile Island? The answer, of course, was to provide a highly trained, full-time monitoring staff, but this raised costs to unacceptable levels.

Usually a dealer for the equipment builder was asked to provide contract service at a reasonable cost; and sometimes this worked well. If the service agency was large enough and had enough experienced men to handle on-site maintenance and breakdowns, there were no problems except from the reluctant building owners who were more financially than technically oriented. Smaller agencies, often with no more skill than needed to repair an engine in a forklift, were loath to enter into service contracts on elaborate equipment and were unable to perform when they did. The same human and business problems applied to the need for emergency-service people seven days a week regardless of holidays, vacations, or illness. When everything went black, usually on Christmas Eve or the like, someone had to be there pronto to get power back on the line. And, of course, no one really wanted to own and pay taxes on an inventory of expensive parts, many of which might never be used but had to be on hand.

If the space I have devoted to these historical problems seems excessive, try for a moment to relate them to the inevitable problems of scattered power sources feeding into a grid. The challenges will be the same—manpower, parts, technical skills, and a reliable organization to make it all work.

For a happier example and a more encouraging perspective, let me return to another scattered-site energy application that even more closely resembles our proposed distributed modular-power projections—oil-field gathering operations. In this industry many, often several hundred, engine-powered pumps ran day and night, year in and year out, pumping from individual wells and delivering crude oil through pipelines to storage tanks far away. The resemblance, if you substitute electricity for crude oil, is striking.

Obviously, each one of these pumping units had absolute requirements for monitoring, lubrication, repairs, replenishment of oil and coolant, and minor service such as spark plug and ignition component replacement. Be assured that although the specific parts may differ with various forms of modular power, each will present essentially the same pattern of demands. Fortunately, in the oil fields certain inherent factors resulted in extremely successful operation.

The nature of the oil industry ensured huge financial and organizational support. This closely resembles the existing utility industry.

- The very large number of pumping sites offered economy of scale. The costs were high but the cost per unit per hour of operation was low.
- Oil operations tended to be within large geographical or political entities and were not often hindered by local political concerns.
- There were no activists lacking technical knowledge to introduce pressures unrelated to the job at hand.

It is worthwhile, therefore, to examine just how these oil-field successes might be applied to modular power. Like a utility with strategically located service bases, each handling a prescribed area, the service agency through a chain of subdealerships was able to provide localized manpower and equipment. Actually, few of the major oil companies elected to perform their own service but preferred to contract for it because their expertise was in other quarters, such as geology, drilling, and refining.

In practice the format consisted of a fleet of trucks with one or two men in each. The crew would arrive at a gathering pump and dispense oil, coolant, and grease through pressure hoses with the speed of a fire-fighting crew. The trucks had storage for high-mortality parts along with needed special tools. Minor parts could be popped off and new ones installed. Those parts needing repairs and rebuilding were brought in each night to central shops, and after being repaired they were consigned to a rotating inventory. Often such replacements were made on the basis of operating hours rather than because of failure. This was less costly than losing well production for twenty-four hours. If oil production can be compared to generating electricity and feeding it into the line for a profit, there is really little difference between what I have described here and what we are considering for the future. With such a plan, widely distributed modular power might be workable.

Attractive as it seems, we should closely examine the factors that made the oil-field system successful.

- The service organization was very large, well financed, and centrally controlled by a single management entity with central records.

HOW PAST EXPERIENCE CAN GUIDE US

- Great care was taken to achieve maximum standardization of equipment, and diversity of equipment manufacturers was minimal.
- The equipment was not new, novel, or in an evolutionary state, although minor engineering improvements were phased in constantly.
- The personnel was nonunion. Any man on the crew could make mechanical, electrical, piping, or other repairs.
- The operation was profitable for both the service organization and the oil companies.

Every one of these factors is important, but putting them all together to obtain the same success with modular power will probably be much more difficult than it was for the oil industry. Some of the reasons are:

- The relationships of the oil companies and their field agencies were internal and consumers were not exposed to or concerned by it. There were no grassroots forces involved as there are with modular power.
- An inherent mental bent characterizing some of the more vocal groups promoting modular power militates against large organizations such as utilities.
- Utilities are to a large degree regulated by government and always are politically sensitive. It would be difficult to establish a working central control in the freewheeling oil-field pattern.
- With expensive equipment and parts to be purchased and jobs to be made available, we could not hope that local political, commercial, and specialized self-interest pressures could be bypassed. There would be trade-offs, perhaps land rights or zoning laws for buying equipment made in the local community.

On the other hand, two factors peculiar to modular power—land use and primitive hardware and technology—introduce unique problems not encountered with energy sources of mature design and high energy density.

Most alternate energy sources require huge land areas because of inherently low power density compared with even a simple diesel engine. Photovoltaic would require arrays over very large areas and probably many such arrays strategically placed so that some would receive solar exposure when others were under clouds. Hydro plants of significant output usually require dams, ponds, and spillways that are land using and often environmentally unacceptable. Wind power seems especially haunted by being location-sensitive relative to wind availability and monstrously oversized relative to output.

One of the latest and best wind generators I've seen has a blade diameter of over thirty feet, sits atop a huge mast, and is complex and costly. Yet it produces only 8 kW under favorable wind conditions. When we stop to think that 8 kW is probably less than half the power demand of even an average single-family house at many times (without electric heating), we can only picture the enormous areas needed for modest power contributions.

Thus, modular power seems to be inherently tied to land acquisition for siting, access roads, and power distribution lines, all of which have political implications. In addition, service travel time and maintaining access roads through blizzards, floods, and ice storms would be a formidable problem.

Returning to the primitive technology and hardware we now know to be a common denominator of most alternative power sources, considerable problems can emerge in both initial procurement and later servicing of equipment. Past experience strongly suggests that all power source equipment would be nearly identical, right down to make and model number. Switch gear and controls would be standardized, and all this would minimize parts inventories, special tools, and record-keeping chores. Standardization would also simplify manpower training and go far to speed and promote efficient service repairs and troubleshooting. The airlines and the military have long recognized that few men are so gifted that they can handle the technical problems of widely diverse equipment efficiently. Our review of serving the needs of oil-field equipment emphasized how the engines and pumps were mature devices that had been thoroughly standardized over the years. Such parts as clutches, controls, exhaust water recovery devices, oil-level equalizers, valves, and delivery controls were off-the-shelf items. Modular power as a conglomerate of small hydro, photovoltaic, solar engine, wind, and other prime movers presents the impossibility of standardizing either hardware or skills.

Even worse, none of these is a mature technology. No matter how we try to lock onto certain equipment, once a technology catches on the technical changes will cascade, making planning very difficult. If we go too far in demanding standardization and opt for long-term investment with currently available state-of-the-art equipment, progress will be impeded. If we wait for the ultimate it will probably never come because progress results only from actual field experience and incremental improvements. Moreover, given a public temper that expects every device to work perfectly the first time and be as reliable as all that has gone before, plus a press that revels in "exposing" public engineering ineptitude and politicians who make hay out of problems they know nothing about, the procurement and selection of modular power equipment will not be an enviable job.

Those of us who have been close to such problems in other fields less exposed to the public eye know all too well that mistakes are made and that it often takes years to "debug" new equipment. In most cases the latter can be done; in a few someone has to swallow hard and simply write off the costs and be guided in the future. But these are private, in-house experiences. Modular power is more likely to make its inevitable errors in the limelight.

Again, looking for historical precedents, we might consider the airline industry, also saddled with a fast-expanding technology during its growing years. Aircraft selection, which might be taken as a sort of model procedure for modular power, has generally been based on strong engineering analysis coupled with judicious looks at the future. Aircraft have always been subject to constant modifications and most undergo several before they even see service. With proper initial design this works well, and if modular power equipment could be designed and selected in this manner, many growing pains might be eased. In addition, realistic terms for phasing out old equipment and phasing in new, with as much compatibility as possible of peripheral gear, controls, base structure and distribution systems, and service techniques must be recognized as essential.

None of these problems is insuperable, and distributed modular power can be made to work if we provide the proper environment for it. I have tried to make the point that a very strong central hand will be needed right from the design and procurement stages, through construction and initial trials, and ever afterward in service and maintenance if costly and damning failures are to be avoided. The siting of each unit must be considered in the light of future maintenance; the designs must reflect plans for modification and updating; the components must be such as to permit ready servicing and rebuilding; and the determination and authority must be present to resist commercial and political pressures, which are certain to be enormous.

Contributors

Howard J. Brown is President of RPM Systems, Inc., a resource planning and management firm in New Haven, Connecticut. He is also Lecturer in Resource Planning at the College of Science in Society at Wesleyan University, in Middletown, Connecticut. Brown was Lecturer at Yale University's School of Forestry and Environmental Studies and its Department of City Planning from 1972 to 1975, and a research associate at Yale's Institute for Social and Policy Studies. He has directed research and education programs under grants from numerous foundations, has served as a resource planning consultant to public agencies from the local to the international level, and to numerous corporations.

Tom Richard Strumolo is President of Energy General, Inc., of Avon, Connecticut, an energy consulting firm specializing in building energy audits and analysis, and in alternative fuels. Before forming Energy General, Strumolo was an energy analyst for the State of Connecticut Energy Department (1974–77), a research associate at Earth Metabolic Design, Inc., and a private consultant.

James W. Benson is Director of the Institute for Ecological Policies in Fairfax, Virginia. From 1977 through 1979 he was project manager of the *Jobs and Energy* study for the Council on Economic Priorities in New York. An energy and political adviser to local, state, and federal officials, Benson travels widely, lecturing at consumer, civic, and educational institutions.

Lisa Frantzis is an energy analyst in the Engineering Sciences Division of Arthur D. Little, Inc. Her work there has included alternative energy technology assessments under contracts with the U.S. Department of Energy, the Solar Energy Research Institute, and the Electric Power Research Institute.

David A. Huettner is Professor of Economics at the University of Oklahoma. He has served as consultant to many public and private institutions including the Office of Naval Research, the Michigan Public Service Commission, the

Office of Technological Assessment of the U.S. Congress, the Oklahoma Department of Energy, the Connecticut Power Facilities Evaluation Council, and numerous corporations. He has written two books and dozens of papers, reports, and articles on energy and economic analysis.

James Gustave Kahn is an energy researcher and consultant and the author of technical papers on such topics as solid waste energy conversion, solar heating, and electric utilities forecasts.

E. F. Lindsley is the engineering editor of *Popular Science* magazine, where he has served since 1970. A lecturer and writer, Lindsley has more than forty years experience in the application of power generation and energy systems throughout the United States.

Amory B. Lovins is a consultant physicist, Policy Advisor to Friends of the Earth, Inc., and Director of Research at Rocky Mountain Institute (Old Snowmass, Colorado). He and his wife and colleague, Hunter, work as a team on energy and resource strategy in more than fifteen countries. They have recently been Luce Visiting Professors at Dartmouth College and received a Mitchell Prize for their work in reallocating electric utilities' capital. Mr. Lovins, a former Oxford don, holds four honorary doctorates; served in 1980–81 on the U.S. Department of Energy's senior advisory board; and has published a dozen books and more than one hundred papers and articles.

David Morris is President of the Institute for Local Self-Reliance in Washington, D.C. He holds degrees in industrial and labor relations, political science, and urban planning. A consultant to federal, state, and local governments on community development, he lectures extensively on the subject of local self-reliance. Morris is a regular columnist for *Solar Age* magazine, has written for many national journals and newspapers, and is the author of several texts on energy and self-reliance. His two most recent books are *Self-Reliant Cities: Energy and the Transformation of Urban America* (Sierra Club Books, 1982) and *The New City States* (Institute for Local Self-Reliance, 1982).

Robert D. Morris is pursuing his doctorate in urban and environmental systems at the University of Wisconsin in Milwaukee. He was a consultant to Milwaukee County in 1981 and a systems engineer at Windworks, Inc., designers and manufacturers of solar and wind electric systems, from 1979 to 1981. He has served as engineer and technical consultant on such projects as an energy planning system for the National Park Service, the design of commercial energy audit procedures, and the development of computer software for use in alternative electrical systems.

Bent Sørensen is Professor of Physics at Roskilde University Center in Denmark. He has worked in theoretical nuclear physics and, since 1971, in energy and environmental physics. About one hundred scientific articles of his have been published, including research monographs entitled "Renewable Energy" (1979) and "Fundamentals of Energy Storage" (with J. Jensen, 1983).

Edward Thompson III is Assistant Professor of Political Science at the University of Louisville, in Kentucky. He received his doctorate from Howard University, where his major fields were American government and public policy. Previously he was Co-Director of the Energy-Equity Task Force at the Environmental Action Foundation, Washington, D.C., as well as a consultant, editor, and author of numerous papers on government and public affairs.

Robert E. Witholder is a Senior Systems Engineer with Teledyne Brown Engineering Company in Huntsville, Alabama, who make payloads for the Marshall Space Flight Center. Prior to joining Teledyne Brown in January 1982, he was an Engineer in Market Research at the Solar Energy Research Institute in Golden, Colorado, from 1977 to 1981; before that he did design and data analysis for NASA on such projects as Apollo and Skylab.

Index

Alternative technologies. See Renewable resources and Individual resources defined
Antitechnology, 32
Aron, Joan B., 199, 200
Autarchy, 24

Bay Area Rapid Transit, 32–33, 244, 247
Biomass energy conversion, 28, 46, 105, 116, 117
Bridsill, Holly, 38
Bullard, Clark, 26

Centralized electric utilities industry: problems of, 3, 13, 21, 183, 216–17; nature of in U.S., 5, 6, 7; as natural monopoly, 7, 39, 184, 215, 220; regulation of, 7, 41, 42, 184–85, 200–05; forecasting future of, 8–13; alternative approaches to, 13–18, 27, 28; economies of scale, 22, 219, 220, 222, 224–25; inflexibility of, 23; and the consumer, 26; and move toward decentralization, 28, 46, 47, 93; and the military, 32–33, 239; history of, 37, 38, 40–42; early competition, 38, 39; role of city government in, 39; v. municipal systems, 43, 44; reliability of, 94, 95; financing alternatives, 132, 134; a program for reform, 185–87, 188–90, 191–95; and penalty for black-/brown-outs, 189; and federal-state policy, 199, 200, 207; traditional economic theory of, 217–18; and unemployment, 229, 230; future choices for, 237–40
Cogeneration, 14, 15; efficient use of, 46, 83, 84; working examples of, 48, 49, 50, 51; and government policy, 202, 203; rates, 202–05; integration with existing grid, 205–07
Committee on Nuclear and Alternative Energy Systems, 78

Conservation of nonrenewable resources: and the consumer, 6, 7, 8; efficacy of, 93, 234, 235, 236
Council on Environmental Quality, 68, 78, 88

Darling, Fraser: *West Highland Survey: An Essay in Human Ecology*, 32
Decentralized electric utilities: and alternative resources, 14, 15, 16; problems of, 17, 111; combining technologies in, 28, 29, 30, 31, 111; and social change, 31, 32; historical precedent for, 37, 38; role of grid in, 109, 110; power transmission in, 110–11; regulation of mixed systems, 112–13; and cogeneration, 113–14; economic assessment of, 118–20; a jobs and energy study, 230–35; ideological interpretations in, 239–41; a historical perspective of, 243–47; factors essential to efficiency of, 247–49; future of, 252–53. See also Renewable resources and Individual resources defined
Department of Energy, 30, 53, 54, 64, 70, 73, 83
Dyson, Freeman, 25

Edison, Thomas, 37–39
Electric streetcar, 38
End-use management, 15
Energy Security Act, 207
Environmental Protection Agency, 81

Federal Energy Regulatory Commission, 43, 48; future role of, 187, 188, 192, 193, 194, 195; and rates for small power producers, 203, 206
Feinberg, Gerald, 3
Forecasting future energy needs, 9–13, 62
Fossil fuels, 20, 93, 105, 106, 109, 183. See also Nonrenewable resources

Gedser mill, 33
Goldberger, Marvin, 19

INDEX

Grid. *See* Utilities grid

Henderson, Hazel: *Creating Alternative Futures*, 4
Hydropower, 47, 48, 70; capacity of, 72–73; future use of, 116

Indigenous resources, 13, 14–15. *See also* Individual resources defined
Individual resources defined: nuclear resources, 3; biomass energy conversion, 28, 105; solid waste conversion, 48; photovoltaics, 51; wind energy, 64; hydropower, 70, 72–73; solar thermal energy conversion, 73; ocean thermal energy conversion, 78; solar energy, 129; fossil fuels, 183–84
Insolation, 55, 129, 139, 147, 158
Insull, Samuel, 39, 40, 55
Investor-owned utilities, 134

Keynesian economics, 6

Load management, 15, 101–02, 103
Load shedding, 104–05
Lotka's principle, 5

Mass transit (electrified), 27
Municipal ownership of utilities, 39, 43–45, 132, 134

National Energy Act of 1978, 201
Natural Gas Act of 1938, 187
Neighborhood electric cooperatives, 38, 55
Nonrenewable resources, 3, 4, 15; traditional policy regarding, 20, 21; and integration with other systems, 86, 118; exhaustion of, 93, 95, 183–84; transition from, 105–06; and environmental concerns, 184
Nuclear power, 93, 95, 105, 109, 234
Nuclear resources, 3, 8, 20, 44, 234, 235, 236

Ocean thermal energy conversion, 78, 81
Odum, Howard, 3–4, 5
Oil embargo of 1973–74, 11, 45
Oligopoly, 24

Photovoltaics: uses, 51, 96; in cities, 52, 53; reliability of, 54, 55, 115; future possibilities of, 55, 56, 70, 117; capacity of, 68–69, 129; integration with existing grid, 97, 98; problems of, 116–17, 251. *See also* Solar energy
Power plants, 7, 21; problems of large plants, 11, 22, 41, 42, 46; location of, 188–91
Private ownership of utilities, 49, 50

Public Utilities Regulatory Policies Act of 1978, 47, 99, 199; Title I of, 201–02; Title II of, 202–03; constitutionality of, 207–08, 209n12; and the environment, 210n22
Putnam wind generator, 243–44

Renewable resources, 15; reliability of, 23, 95, 100–01, 106, 163, 174–76, 192–94; appropriate use of, 29, 30, 46, 47; issues regarding, 61, 62; background, 63, 64; impact on energy supply, 87–89; integration with existing grid, 98–99, 100–04, 105, 106, 164–68; load management of, 101–02, 103, 104–05; storage of, 103; and capacity credit, 117–18; hybrid systems of, 167, 172, 174, 176, 178, 179, 180n11; and conservation, 236; and problems of land use, 251, 252
Rural electric cooperatives, 134

Sandusky wind machine, 33
Socolow, Rob: *Patient Earth*, 26
Solar energy, 28, 46; reliability of, 101, 129, 136–40; integration with existing grid, 112; economics of, 121–23; value to user of, 124–26; a methodology to derive value of, 126, 127, 128; a regional analysis of, 147, 152, 153–55, 157–59
Solar thermal energy conversion, 73, 75, 77, 78, 129, 140
Solid waste conversion, 48, 81, 82
Sprague, Frank, 38
Storage of energy, 17, 18, 95, 103; technologies for, 96, 105, 114–15; expense of, 97, 99; and biomass conversion, 116; in hybrid systems, 176, 177

Technocracy, 26
Technology: high v. low, 32, 33–35
Tennessee Valley Authority, 132
Three Mile Island nuclear plant, 247, 249
Turbines: steam, 39, 40, 83; wind, 46, 98, 123, 129, 132; gas, 83, 84, 113

U.S. Army Corps of Engineers, 70, 246
Uranium, 3, 183
Utilities grid, 14, 16; defined, 94; and integration of alternative technologies, 98, 99; use of, 109–12, 118
Utility financing, 132, 134

Wilcox, Delos, 41
Williams, Robert H.: *Cogeneration: An Assessment of Commercial Readiness*, 83
Wind energy, 17, 28, 46; integration into existing grid, 64; reliability of, 64, 66, 101, 129; future use of, 68, 116, 117; problems of, 251
Wright, Deil S., 199, 200

RAYMOND H. FOGLER LIBRARY

DATE DUE

ARE SUBJECT TO
TWO

BOOKS
RECALL AFTER

FEB 10 1988
MAY 27 1988
AUG 1988